WAR: A CRUEL NECESSITY?

WAR: A CRUEL NECESSITY?

The Bases of Institutionalized Violence

Edited by
ROBERT A. HINDE and
HELEN E. WATSON

Tauris Academic Studies
I. B. Tauris Publishers · London · New York

Published in 1995 by
Tauris Academic Studies
An imprint of I.B.Tauris and Co Ltd
45 Bloomsbury Square
London WC1A 2HY

175 Fifth Avenue
New York
NY 10010

In the United States of America
and Canada distributed by
St Martin's Press
175 Fifth Avenue
New York
NY 10010

A full CIP record for this book is available from the British Library

Library of Congress catalog card number: 94–60184
A full CIP record is available from the Library of Congress

ISBN 1–85043–824–2

Copy-edited and laser-set by Selro Publishing Services, Oxford
Printed and bound in Great Britain by WBC Ltd, Bridgend, Mid Glamorgan

Contents

Acronyms

AKEL	Anorthotiko Komma Ergazomenon Laon (Greek Cypriot Communist Party)
AWAC	Airborne warning and control systems aircraft
AWOL	Absent without order of leave
CIA	Central Intelligence Agency (USA)
COCOM	Coordinating Committee on East–West Trade
CPP	Congress People's Party (Ghana)
CSCE	Conference on Security and Cooperation in Europe
EC	European Community
EEC	European Economic Community
EMU	Education for Mutual Understanding
EOKA	Ethniki Organosis Kipriakon Agonos (National Organization of Cypriot Struggle)
EOKA B	As above, but with the B added after its unsuccessful resurrection in 1971
ESRC	Economic and Social Research Council
FHAO	Facing History and Ourselves
HQ	headquarters
IMF	International Monetary Fund
IRA	Irish Republican Army
ISIS	International Security Information Services
MI6	Military Intelligence (Espionage Department)
MoD	Ministry of Defence
MP	Member of Parliament
MRC	Medical Research Council
NATO	North Atlantic Treaty Organization
NIN	Serbian weekly published in Belgrade
OPEC	Organization of Petroleum Exporting Countries
PNG	Papua New Guinea
POWs	Prisoners of War
PR	public relations

SS	*Schutzstaffel* (Nazi elite corps)
TLB	Top Level Budget
TMT	Turk Mukavemet Teshkilati (Turkish Cypriot equivalent of EOKA)
UBP	Ulusal Birliyi Partisi (Turkish Cypriot party)
UFF	Ulster Freedom Fighters (Unionist paramilitary)
UK	United Kingdom
UN	United Nations
UNGA	United Nations General Assembly
UNHCR	United Nations High Commission for Refugees
UNESCO	United Nations Educational, Scientific, and Cultural Organization
US	United States
USA	United States of America
WEU	Western European Union

Contributors

Michael Brzoska. Lecturer at the University of Hamburg, Germany, and researcher at the Unit for the Study of Wars, Armaments and Development. Formerly at the Stockholm International Peace Research Institute. Author of numerous articles on arms control and related topics.

Susan Drucker-Brown. ESRC fellow at the Department of Social Anthropology, University of Cambridge. Has worked on the Kingdom of Mamprusi in North Ghana and is extending her work to a cross-cultural study of war.

Scilla Elworthy. Director of the Oxford Research Group, editor of *How Nuclear Weapons Decisions are Made* and author of numerous articles on decision-making responsibility for weapons and related topics.

Seymour Feshbach. Professor of psychology at the University of California, Los Angeles. Has devoted most of his research to the study of aggression and the development of aggressiveness. In recent years has been writing on the relations between, and bases of, patriotism and nationalism.

Jo Groebel. Professor in the Department of Social Psychology and Mass Communication, University of Utrecht, the Netherlands, formerly at University of Landau, West Germany. Has worked on the psychology of terrorism, and on the psychology of the media.

Robert Hinde. Former master of St John's College, Cambridge, and formerly Royal Society research professor and honorary director, MRC Unit on the Development and Integration of Behaviour. Author of *Individuals, Relationships and Culture* and *The Institution of War* and co-editor of *Aggression and War.*

Gilbert Lewis. Lecturer in the Department of Social Anthropology and fellow of St John's College, Cambridge. Has worked extensively in Papua New Guinea. Publications include *Knowledge of Illness in a Sepik Society* and *The Day of Shining Red.*

Mark Edward Lewis. Lecturer in the Department of Oriental Studies, University of Cambridge. Author of *Sanctioned Violence in Early China.*

Joseph McDermott. Lecturer in the Faculty of Oriental Studies and fellow of St John's College, Cambridge. Primarily a Sinologist, he is also interested in Japanese history, especially Sino–Japanese relations. At present he is editing a book on ritual at the Chinese court.

George Mosse. Former professor of history at the University of Wisconsin, Madison, and the Hebrew University, Jerusalem; A. D. White professor at large, Cornell University, sometime visiting fellow, Pembroke College, Cambridge. Author of numerous books on war-related issues, including *The Crisis of German Ideology*; *Fallen Soldiers*; *The Nationalization of the Masses*; *Nationalism and Sexuality* and *Toward the Final Solution.*

Dominic Murray. Professor of peace studies, University of Limerick. Main research interests are sectarianism and the community divide and segregated education in Northern Ireland. Author of *Worlds Apart: Segregated Schools in Northern Ireland.*

Robert Neild. Emeritus professor of economics and fellow of Trinity College, Cambridge. Author of numerous books and articles on economics, strategy and disarmament.

Yiannis Papadakis. Fellow of Churchill College and member of the Department of Social Anthropology, University of Cambridge. At present completing a study of nationalism in Cyprus.

Linda Siegel. Professor at Ontario Institute for Studies in Education, Canada. A developmental psychologist who has worked extensively on socialization with special reference to cooperative and aggressive behaviour.

Sharon Smith. Postgraduate student at the Ontario Institute for Studies in Education, Canada. She has conducted research in developmental and military psychology.

Cornelia Sorabji. Research fellow of the British Academy and lecturer at the School of Cultural and Community Studies, University of Sussex. Her main research interests are religion and identity, fundamentalist Islam and Yugoslavia (Bosnia).

Hylke Tromp. Professor at the University of Gröningen, the Netherlands and behavioural scientist spanning several disciplines. Is a member of Pugwash and has written on the nature of militarism.

Helen Watson. Lecturer at the Department of Social Anthropology and fellow of St John's College, Cambridge. Interested in religion and culture. Has worked in North Africa and Europe (France and Ireland). Publications include a recent book on the women in the City of the Dead — based on a study carried out near Cairo. Currently studying the bases of the troubles in Northern Ireland.

Acknowledgements

The chapters in this volume are based on papers presented at a conference held at St John's College, Cambridge, from 4 to 8 December 1993. The conference was supported by the Carnegie Corporation of New York and the John D. and Catherine T. MacArthur Foundation of Chicago; it was sponsored by UNESCO. We are grateful to Margaret Hauser for her meticulous help in editing the manuscript.

General Introduction: War and the Institution of War

After his execution, the body of King Charles I was placed in the Banqueting House in Whitehall. According to an often-cited tradition, at 2.00 o'clock in the morning a heavily disguised Oliver Cromwell inspected the body closely, shook his head and sighed, 'Cruel necessity.' But is war necessary? Can some conflicts be solved only by violence?

Human history is indeed often presented as primarily a history of wars and battles, conquests and defeats. While that is only one perspective amongst many possible ones, violence of one sort or another has certainly been, if not centre-stage, at least lurking in the wings throughout the human story. Man (especially Man, but also Woman) clearly has the propensity not only to behave aggressively to other humans but also to do so in an organized way and not infrequently with calculated cruelty.

But this does not mean that war is inevitable (Groebel and Hinde 1989a); some societies maintain long periods of peace (Howell and Willis 1989). While the capacity to wage war is a near-ubiquitous outcome of human development, war itself is not in our genes, but only the potential for war. Furthermore the capacity to behave with compassion, to show altruism, to cooperate for the common good, are also near-ubiquitous outcomes of human development (Hinde and Groebel 1991). Humans may be competitive or cooperative, violent or compassionate, aggressive or prosocial.

Should history therefore be presented as an oscillation between the two with wars interrupting periods of peace? It is not so simple as that. War, however destructive it may be, involves cooperation and self-sacrifice, and periods of peace involve conflict and rivalry. Perhaps we should therefore seek to identify the factors that cause peacetime rivalry to tip over into socially destructive war (cf. Singer 1989). Here, however, general answers are difficult if not impossible to find, because every war has multiple 'causes' and the spectrum of interacting factors involved differs from case to case.

There is, however, an underlying issue. The destructiveness of war is abhorrent to most people, and especially to those involved. How is it, then, that war comes to be seen as an acceptable means of settling conflicts of interest? How is it that leaders of nations or groups commit their followers to armed conflict, knowing the risks involved? Do individuals go willingly, even gladly, to battle knowing they may not return?

Such questions as these are the ultimate concern of the contributors to this volume. However, they seem to reduce the problem to matters solely of individual motivation, and therein lies a danger. Because attempts by one individual to harm another and attacks by one nation on another are both described as involving aggression, it is often supposed that war depends on the aggressiveness of individuals. In fact, the role of individual aggressiveness in war is both more complex and in many ways less important than such a simple-minded approach indicates. There are also the further issues of how behaviour in situations of conflict should be interpreted, and the dangers of considering institutionalized violence as a monolithic category. For instance, observers and combatants may perceive the nature and goals of aggression in war in very different terms. And how relevant is the notion of aggression for those engaged in violent acts that are perceived to be part of a liberation struggle? These concerns raise questions about the analytical use of a general notion of aggression without consideration of the specific historical and cultural context. They also require some digression.

It is necessary first to distinguish several interrelated questions — the role of individual aggressiveness in making war seem acceptable, its role in causing particular wars, its role in motivating the participants in a war, and the reciprocal issue of the effect of a state of war on individual aggressiveness. It is also useful to distinguish 'impulsive' aggression, which has the sole purpose of inflicting harm on the opponent as victim, from instrumental aggression, which has the further goal of obtaining access to some scarce resource. (That such distinctions, though heuristically useful, are not absolute is not a matter that need detain us here.)

While there have no doubt been cases where interstate wars were started through the desire of one leader to destroy another, or where leaders actively sought to cause others suffering (Frank 1994), impulsive aggression cannot be considered as an important cause of wars. It may be an important factor in ethnic or religious wars, but even there is probably sanctioned by the existing state of war rather than causing the latter. The impulsive aggression shown in the massacre at My Lei in the Vietnam War is to be seen as a product of the state of war, and not as its cause; and the atrocities following the break-up of the former Yugoslavia, discussed by Sorabji and Tromp (Chapters 6 and 8), occurred primarily once the state of war existed. As shown by Sorabji, individual aggression, though prominent, was sanctioned and even encouraged by political leaders in the war for their own ends. Even in pre-state warfare, impulsive aggression plays at most a minor direct role in motivating combatants (McCauley 1990).

However, 'impulsive' aggression may play an indirect role. As we shall see, the propaganda used to support both the idea of war and particular wars often involves portraying the enemy as evil and non-human, and thus meet to be destroyed. Terms such as 'infidel' label an enemy as beyond the pale — not merely as one to whom the sanctions against killing do not apply, but as one who should be killed.

However, instrumental aggression, involving access to material resources, such as land, food and trade goods, may be an important cause of pre-state warfare (Chagnon 1990; Ferguson 1990). (Women may be the cause of violence between individuals, but access to women is less often a cause of war (McCauley 1990).) In addition, in pre-state warfare instrumental aggression may be a by-product of successful war, and thus play a part in the motivation of individuals.

Access to material resources also often plays a role in the instigation of modern war — for instance the Pacific war of 1942–5. In that sense, states can be said to show (instrumental) aggression. That, however, involves personalization of the state, and clouds the issue. Governments may be acquisitive or assertive, and use force as a means to an end, but the aggressiveness of individuals can seldom play more than a very minor part in this. And access to material resources plays virtually no role in motivating combatants to inflict harm on their enemies.

The only reservation here is that intangible resources may be important. Individual success in war can bring status, signalled by medals and other honours, or simply by 'glory'. The 'glory' may be that of the ship, regiment or squadron, rather than of the individual. Furthermore, intangible resources may be linked to tangible ones — the first Duke of Marlborough was rewarded for his army's victory at Blenheim by the gift of Blenheim Palace from the British government. From the perspective of an individual engaged in what is perceived to be 'holy war' an individualistic sense of 'personal glory' may be eclipsed by an ideological concern with the collective social good which will be secured by the defeat of evil. These are issues that we shall return to later: the immediate point is that neither the incidence of war nor the behaviour of individuals in modern war can be ascribed simply to 'aggression'. Both involve issues very much more complicated than that.

There are other reasons why the issue of how people come to accept war cannot be understood fully in terms of individual motivation alone. First, individuals are socialized within societies with particular cultures in ways that permit, in certain circumstances, motivation to go to war to become prepotent. Here we must remember that culture is not a given, but is continuously re-created by the actions of individuals in their relationships with each other. It is important to examine the interplay between cultural values and ideals and social reality, as well as the dynamic tension between what is normatively recognized as acceptable behaviour and formal or legal definitions of the same. For instance, the change in the legal definition of rape to include rape within marriage in the Britain of the 1990s both reflected and affected contemporary attitudes towards sexual violence and marital relations. In a similar way, in recent decades the incidence of divorce has affected people's attitudes to divorce as well as vice versa: there is in fact a continuing dialectic between the way in which individuals behave and the way in which individuals see it as proper to behave.

Second, the goals sought in war are at least in part group goals, and do not necessarily materially benefit the individuals who must bear the brunt of the fighting. These group goals may, however, be internalized by individuals, perhaps as a consequence of deliberate manipulation of cultural values by leaders.

And third, at least in modern war, the immediate goals of individuals do not necessarily involve fighting, but may include the making of munitions, the care of the wounded, the distribution of resources (as in rationing) and so on.

Modern war, at least, must therefore be seen as an institution, with numerous constituent roles. Each role has associated with it specific rights and duties, and it is these duties that are the immediate goal-setters for individuals. The question posed above then becomes, 'What are the forces that maintain bellicose values, and in particular the institution of war, shaping its constituent roles so that the incumbents will go willingly to war?' This is a question distinct from that of the causes of war and, it may be suggested, in a certain sense more fundamental in that it is general to all violent conflicts, even to pre-state war where the number of distinct roles is very small.

Thus the questions posed above focus on the issue of the forces that maintain the institution of war. These are diverse and range from pervasive cultural issues, including the language we use, the books we read, and the religion by which we are influenced, to the various sub-institutions that support the institution of war, including the armed services and the munitions industry. An initial exploration of some of these issues was written with the cold war era in the background (Hinde (ed.) 1991; see also Hinde 1993). Here the problems are pursued in somewhat greater depth, with a greater focus on the religious and ethnic conflicts that have dampened the hopes that were aroused by the breaking of the Berlin Wall.

The chapters fall into three sections. The first contains six examples of warfare involving violence that lasted, continuously or intermittently, over a long period. The second is concerned with some of the pervasive cultural factors that support the acceptability of war. Finally, the third section focuses on some historical and economic issues and on one of the sub-institutions of the institution of war — the arms industry.

Part I

PERSISTING VIOLENCE

Introduction

All the authors in this volume agree that support for the institution of war involves forces beyond the aggressive motivation of individuals. That immediately raises the question of what is meant by war. In Chapter 8 Tromp emphasizes how the nature of war has changed over the centuries, making definition difficult. But even looking at current conflicts we must recognize that the real world cannot always be fitted into the pigeon-holes beloved of academics, and there is bound to be a penumbra of cases about which opinions differ. However, certain issues are clear. First, war involves attempts to inflict physical injury on the enemy: economic sanctions, competition, threat and bombast do not involve war in this sense of the term. Second, war takes place between groups whose individual members perceive the group boundaries and the group differences as distinct and important. It involves socially sanctioned injury of out-group members. Third, war involves at least a degree of centralized organization and collective mobilization of individuals, and is thus clearly distinct from, say, mob violence. Fourth, although a war may be decided in a brief battle, the state of war usually has considerable duration.

The global scale of the 1914–18 and 1939–45 world wars gave them a paradigmatic status as exemplars of war. But in a sense they are the exceptions: most wars are much more local in scope and more intermittent in intensity, while often longer in total duration. As G. Lewis (Chapter 2) emphasizes, the scale of the conflict and especially the numbers of individuals actively involved affect its nature and the forces that maintain it. In some ways it is long-drawn-out conflicts, in which periods of relatively peaceful coexistence are punctuated by socially destructive war, that exemplify even better than the world wars the forces that make war acceptable. In this section some examples of such conflicts are discussed.

The contributions here range from tribal conflicts to modern war, and involve issues of natural resources, ethnicity and religion. What is surprising, therefore, is the number of common themes that emerge. These are of two sorts — those concerned with the occurrence of war

Introduction

All the authors in this volume agree that support for the institution of war involves forces beyond the aggressive motivation of individuals. That immediately raises the question of what is meant by war. In Chapter 8 Tromp emphasizes how the nature of war has changed over the centuries, making definition difficult. But even looking at current conflicts we must recognize that the real world cannot always be fitted into the pigeon-holes beloved of academics, and there is bound to be a penumbra of cases about which opinions differ. However, certain issues are clear. First, war involves attempts to inflict physical injury on the enemy: economic sanctions, competition, threat and bombast do not involve war in this sense of the term. Second, war takes place between groups whose individual members perceive the group boundaries and the group differences as distinct and important. It involves socially sanctioned injury of out-group members. Third, war involves at least a degree of centralized organization and collective mobilization of individuals, and is thus clearly distinct from, say, mob violence. Fourth, although a war may be decided in a brief battle, the state of war usually has considerable duration.

The global scale of the 1914–18 and 1939–45 world wars gave them a paradigmatic status as exemplars of war. But in a sense they are the exceptions: most wars are much more local in scope and more intermittent in intensity, while often longer in total duration. As G. Lewis (Chapter 2) emphasizes, the scale of the conflict and especially the numbers of individuals actively involved affect its nature and the forces that maintain it. In some ways it is long-drawn-out conflicts, in which periods of relatively peaceful coexistence are punctuated by socially destructive war, that exemplify even better than the world wars the forces that make war acceptable. In this section some examples of such conflicts are discussed.

The contributions here range from tribal conflicts to modern war, and involve issues of natural resources, ethnicity and religion. What is surprising, therefore, is the number of common themes that emerge. These are of two sorts — those concerned with the occurrence of war and the justification of the use of violence, and those concerned with the motivation of participants. Of course, there is an interplay between these two sets of issues, but it is convenient here to focus on the former, and postpone the latter to Part II. Here, then, are some of the issues that come up in this section.

1. Religious and/or ethnic issues are paramount in all these cases. While East–West nuclear war was an overriding fear, the potential disruptiveness of ethnic and religious issues was less apparent, but now they are clearly more important than East–West rivalry or the economic competition that has often been important in the past (see, for example, Connor 1994).
2. The importance of distinguishing between the situation as perceived by those involved in it and the situation as perceived by outsiders is emphasized especially by Papadakis (Chapter 4) with respect to Cyprus, by Murray (Chapter 5) over the conflict in Northern Ireland, and by Sorabji (Chapter 6) with respect to Bosnia-Hercegovina. Furthermore, the meaning of aspects of the conflict or of events in its course may be multiple and complex and differ according to the perspective of the individual involved (Sorabji). One consequence of this is that violence perceived as quite unacceptable by outsiders may be seen as inevitable and legitimate by those involved.
3. The social knowledge of historical relations between warring parties is seen as a significant issue in all these cases. McDermott (Chapter 13) suggests that the aggressiveness of Japan in the first half of this century was a consequence of the relations between the East Asian states, including a failure to work out a systematic way of dealing with neighbouring states. On a much shorter time scale, G. Lewis (Chapter 2) describes the tribal wars in New Guinea as part of a complex field of institutional relationships involving enmity and alliance, rules of marriage and expectations of gifts, loyalty, secrecy and so on. In other cases it is history as perceived by those involved that is important. Even in tribal wars revenge for supposed past injuries may be crucial for the maintenance of conflict (G. Lewis Chapter 2; Drucker-Brown Chapter 3), and in ancient China some wars were seen as 'just' by virtue of the supposed past wrongs of the enemy (M. Lewis Chapter 1). Of course historical knowledge is a social construct, and conflicting versions of history may be used to explain and justify the positions held by opposing sides in war.

History as perceived by the individuals on one side may be quite different from that perceived by individuals on the other (Papadakis Chapter 6). The versions of history espoused by outsiders, non-combatants and distant observers may be quite different from the historical understanding of the conflict held by any of those directly involved (Murray Chapter 5; Sorabji Chapter 6). Media bias, various forms of propaganda and misinformation contribute to the creation and spread of such contested versions of history over time.

4. In some cases, violence is seen as the only possible means of achieving an important end result. It is an obvious tool in pursuit of the critical goals of status or revenge in tribal wars, but it may be supplemented by other symbolic and non-physical means, such as sorcery (G. Lewis and Drucker-Brown in Chapters 2 and 3). Violence may also be perceived as the only means open to a minority to correct historical injustices imposed by a dominant majority group (Murray Chapter 5). Sorabji (Chapter 6) suggests that the issue of 'alternative options' is irrelevant in the case of Bosnia; the leaders of Slovenia, Serbia and Croatia had no interest in 'seeking other means', seeing war as a way to divide and destroy Bosnia-Hercegovina. The concept of a just war can also make, and be used to make, war more probable: if higher powers demand battle, battle there must be. M. Lewis (Chapter 1) discusses this issue with respect to the Warring States in fifth- to third-century BC China.
5. Wars usually involve at least two sides in the central arena of conflict. However, there are never just two parties involved. It is important to ask what is meant by 'side', who is involved, how and why. On some occasions a war may be seen to involve two sides, made up of a number of armies, but defined by allegiance to a common political goal such as democracy. Conversely, from an outsider's perspective it may be difficult to differentiate the parties involved in war, or to understand how one 'side' perceives itself to be separate from the other. One of the most interesting conclusions drawn by several of these authors concerns the role, deliberate or unwitting, of outsiders. Thus the UK and Irish governments have played a crucial role in Northern Ireland (Murray Chapter 5); the UK, Greece and Turkey in Cyprus (Papadakis Chapter 4); the European Community and the UN in former Yugoslavia (Sorabji Chapter 6); and the central government in Ghana (Drucker-Brown Chapter 3). In Japan, an important factor was the search for external markets (McDermott Chapter 13).

6. In some cases the flames of war were fanned by the self-interest of particular individuals (G. Lewis and Drucker-Brown in Chapters 2 and 3) or by those seeking advantage for the groups they represented (Papadakis, Murray and Sorabji in Chapters 4, 5 and 6), or both. Argentina's invasion of the Falkland Islands seems to have caused a massive swing of popular opinion from near-rebellion to ecstatic support of the junta (Hastings and Jenkins 1983), and Sorabji (Chapter 6) suggests that leaders of the new republics emerging from Yugoslavia needed war to maintain power over their own populations.
7. For war to occur, some degree of initial separation seems to be a major facilitator. In parts of Northern Ireland there is a high degree of demographic separation, with the effect that people may live in predominantly Catholic or Protestant neighbourhoods (Murray, Chapter 5). Where separation is only partial, labels may serve to emphasize difference (Sorabji Chapter 6). Indeed, as these two authors show, it is where separation is only partial that its importance is easiest to discern. And where spatial separation is only partial or absent, indirect discrimination in employment may maintain or exacerbate economic segregation (see also Newsam 1989).
8. Among these authors, it is Sorabji and Tromp (Chapters 6 and 8), writing about the break-up of Yugoslavia, who are particularly concerned with individual aggression. However Sorabji states that this does not imply that those involved were naturally or culturally aggressive, and maintains that the perpetrators of violent acts do not necessarily see themselves as aggressive, but hold rather that violence was forced upon them by circumstances or by superiors. Tromp analyses some of the ways in which callous violence became possible in individuals who under other circumstances might behave with gentleness and compassion.

1. The Warring State in China as Institution and Idea

Mark Edward Lewis

During the Warring States period (ca. 480–221 BC) China was divided into realms based on the total mobilization of their male subjects for military service. All political institutions were devoted to organizing and provisioning the armies (Lewis 1990: Chapter 2). Consequently, writers of the period devoted considerable attention to the question of what caused men to fight for the state. While some suggested that warfare was an expression of a natural human bellicosity, most agreed that it was a socially created form of action and that men had to be *induced* to fight. Thus these writings offer considerable evidence regarding the methods by which rulers and thinkers of the Warring States period justified and perpetuated the institution of war.

This material, however, shows us nothing of the people's experience of military service and their attitudes to combat. For evidence on these questions we must look at later Chinese history, when universal military service had vanished, but the *idea* of a society organized for war survived in various forms. As I will discuss in the second section of this chapter, the survival of the army as a dream and an aspiration in defiance of a government policy of demilitarization has important consequences for our understanding of the underpinnings of warfare.

The social bases of warfare in Warring States China find their most detailed treatment in the writings of the so-called 'legalist' school. These were administrators or scholars who developed political models from the actual practices that had been introduced piecemeal over the preceding centuries. Their guiding maxims were most clearly articulated by the school of Shang Yang (d. 338 BC). According to this school, the state sought to attain wealth and power, and this was achieved through agriculture and warfare. The link between these two activities lay in the fact that the armies of the period were based on mass peasant levies, so that those who practised agriculture also engaged in warfare. Power was won by motivating these men to toil in the fields and risk their lives in battle, so the central question was how to inspire and guide the peasantry.

The 'legalist' answer was to use rewards and punishments. These were, according to Han Fei (d. 233 BC), the 'two handles' by which the ruler wielded power. If rewards for heroism and punishments for cowardice were written in legal codes, rigorously enforced by obedient officials, and in force long enough for people to gain faith in government responses, then the population could be led to fight and die for the ruler.

These theories found their institutional expression in the reforms of Shang Yang in Qin. The entire population was ranked in a hierarchy of titles, and these were awarded solely on the basis of military merit won by presenting the heads or ears of slain enemies. Promotion in the scale of titles meant not only higher status but also allotments of land and slaves, as well as the privilege of surrendering rank to redeem punishments for criminal offences. Evidence from Qin legal documents discovered at Yunmeng shows that this ranking system was indeed put into practice.

Shang Yang also instituted a scale of punishments, ranging from fines through labour service to physical mutilation and death. To secure the zealous enforcement of these punishments he divided the population into units of mutual responsibility comprising five households. The members of these households were liable for the crimes of their fellows in civil life and for their cowardice in combat. Such groups allowed the state to supervise a large population with a relatively small number of officials, and also encouraged steadfastness in battle through feelings of responsibility towards, or fear of, one's closest acquaintances. Basing military recruitment on groupings of neighbours also served to identify the army with the civil populace (Hulsewe 1985: 14–18, 84–5, 145, 185; M. Lewis 1990: 61–5).

Rewards and punishments were, however, only the first step. Once an extended period of rigorous government and military success had created order and prosperity, then the people would come to do their duty without the promise of reward or the threat of punishment. In this second stage the people recognized the state-created order as the source of their blessings and thus fought for the preservation of that order itself, now locating their private benefit in the public realm. The 'legalists', like all writers of the period, saw unity and collective resolve as the ultimate keys to victory, so the ruler's position became secure only when he moved beyond rewarding and punishing individuals to the creation of a unified body politic (Duyvendak 1928: 194).

While the 'legalists' were the greatest exponents of a society organized for war, other philosophical schools also discussed the social bases of military action. The Confucian school emphasized the role of education and the sense of obligation fostered within the family. At the end of the fifth century BC, Confucius outlined a curriculum that included two martial elements: charioteering and archery. He also asserted the necessity of instructing the people before leading them to war, and such instruction included not only specifically martial skills but also a sense of shame. The dread of shame and desire for honour had inspired the combats of the Zhou nobles in the centuries preceding the Warring States, and the conservative social programme of the early Confucians inherited this vision of the social bases of combat (M. Lewis 1990: 36–43).

Warring States Confucians such as Mencius (fourth century BC) and Xunzi (third century BC) carried forward this emphasis on moral education as the basis of warfare. Although in disagreement on many points, these two agreed that the good ruler would triumph through the unanimous support of the people. Good government won the people's loyalty, and this loyalty guaranteed ultimate triumph over less principled rivals. Xunzi also presented a detailed critique of the contemporary practice of relying on rewards and punishments to motivate the army. He argued that such policies appealed entirely to self-interest and encouraged a calculating mentality which in the long run alienated the soldiery from their superiors and from any conception of a public good. Only a sense of selfless devotion and loyalty forged through education and patterned on the idealized feelings of sons for fathers could achieve the complete devotion necessary for victory (Knoblock 1990: 211–34; M. Lewis 1990: 66, 128–30). Xunzi and Mencius were also among the first to enunciate the doctrine of the moral war, in which warfare was justified as a form of controlling the irremediably wicked. This 'just war' doctrine became central to Chinese discussions of warfare in the late Warring States and the empire.

Xunzi's work and related chapters from one of the ritual canons also contain Confucian discussions of training in the form of war dances that ritually re-enacted the Zhou conquest of the Shang. The direct applicability of these passages to Warring States practice is uncertain, but they are supported by textual and archaeological evidence on the use of music to guide men into combat and secure unity in their actions.

Commanders used music to relay commands, to secure motion in unison and to inspire the soldiers to action. Drums, gongs and flutes were all employed for these purposes, and literary accounts depict the use of musical accompaniment in combat. Indeed, an early dictionary stated that the word 'martial' meant 'to dance' and that the advance of an army to the beat of the drum was a great dance (M. Lewis 1990: 226–8, 231, 235).

The use of music in combat indicates the importance of aesthetics to the institution of warfare. A second aesthetic consideration was the visual effect of uniforms and weapons. The splendid appearance of noble warriors was a recurring topic in the classical Odes of the early Zhou (twelfth through seventh century BC), and costumes certainly added much to the mystique of soldiers in the Warring States. This is not discussed in detail in the philosophical texts, but one historical source notes the distinction of status marked by varieties of armour. More importantly, the art of the period clearly reveals a concern with uniforms. The fine costumes and elaborate hairstyles of the terracotta army of Qin Shihuang contribute much to its impact on the viewer, and the living spectacle with coloured fabrics and brightly lacquered armour would have been even more impressive. These statues bespeak a clear awareness of the importance of physical appearance in defining the army as a special group and marking internal divisions of rank and function. The depiction of fierce animals or monsters on shields and armour also contributes to the aesthetic ideal of the Warring States soldier.

A third major philosophical school, the Mohists, wrote extensively on the institution of war. The earlier sections of their canon denounced warfare as a destructive activity whose costs outweighed its gains. However, the Mohists soon replaced this total rejection with a support of defensive wars, and many of them formed roving military units that specialized in siege warfare and hired themselves out to small states seeking assistance against their larger neighbours. These men produced writings on the defence of cities which offer valuable insights into the actual practices by which a population was mobilized for war. These include accounts of registering populations in units of five and ten; using women, children and old men to assist in warfare; taking hostages from important families to secure their loyalty; and distributing rewards and punishments (Yates 1980: 17–34).

The Warring States also produced a body of treatises explicitly devoted to military affairs and their place in the socio-political order.

These texts devote considerable attention to the problems of organizing a society to be militarily effective and of leading men to fight. Several of their basic policies resemble those found in the other philosophical schools, such as distributing rewards and punishments, obtaining the affection of the people through benevolent government, uniting the people through uniformity of laws and administration, invoking the mutual love of kin as a motivation to fight, mobilizing support for war through public denunciations of the enemy's crimes, and alienating the people of hostile states from their rulers through the avoidance of pillage. However, these texts also articulated their own distinctive ideas on the institution of war.

First, while accepting the consensus that successful warfare depends on proper administration, several of the military treatises insist on maintaining clear distinctions between the army and civil society. One text argued that if the virtues of civil society — courtesy, the spirit of yielding, measured speech, modesty — were carried into the army, then the army would be weak. Thus those in the military should not bow, men on watch should not bend and scurry in the manner of courtiers, and soldiers should invert the conventional precedence of age over youth. This sense of the military as an inverted civil society was marked in other texts by a reversal of such basic ritual principles as the priority of left over right.

Another feature of the military treatises is their emphasis on the importance of the commander. For example, they insist that the general in the field cannot be constrained by commands issued from the capital. This was because unity was the key to military success, the soldiers were united only by absolute faith in their commander, and such faith was possible only if the commander was the unchallenged ruler in his own sphere. His power should extend not only to making strategic and tactical decisions but also to controlling the distribution of rewards and punishments in the field, including the power of summary execution and to distribute lands and titles. If the commander were controlled by the ruler, then the army would possess 'two minds' and cease to be an all-powerful 'single body'.

The metaphor of the army as a body with the general as its mind is a recurring trope in the military texts, and it is linked to other ideas regarding the institution of war. First, the treatises insist that the army forms a body as a result of extended training. Only such discipline could overcome the diverse sentiments of the individual soldiers and forge them into cohesive units that acted as one at the behest of their officers. Thus squads of five were the smallest discrete units of military action, while individual courage or cowardice was dismissed as equally damaging to collective endeavour. Here again dance served as the image of coordinated movement, and the drum became the standard attribute of or metonym for the commander. The importance of instilling discipline and *esprit de corps* by unison drills has been noted by William McNeill, and this figured prominently in the Warring States theories of the bases of war (McNeill 1982: 125–39).

The ability of the officers to lead their men into battle did not depend entirely on their control of rewards and punishments, nor on the effects of repeated drill. Military theorists also insisted that commanders must share the material discomforts of their men. They must eat only after their men were fed, and go to sleep only when their men were at rest in their tents. Through this conspicuous display of concern and shared suffering they would appear as fathers to their men and elicit total devotion. This figure of the beloved commander who inspires the adoration of his troops is well known in the West from the time of Alexander down to modern times.

Yet another feature of the military treatises is their frequent appeal to the importance of 'energy' or 'morale', *qi*. This spirit energy was thought to be the basis of all natural dynamism and life, and in human beings it appeared as a propensity to fight. However, this suggestion of a natural bellicosity was repeatedly undercut by the military treatises, which insisted that warfare entailed not the release of this energy but its containment, transformation and manipulation. Texts discuss how the energy of the troops could be aroused, sustained and channelled. They offer advice on 'divining' the spirit of the enemy and depleting their energy through clever tricks. Thus even this idea, which suggested a natural origin of warfare, ultimately insisted on its artificial character and the necessity of manipulating men to fight (Sawyer 1993: 32–4, 116–25, 203–5, 232–7 and passim; M. Lewis 1990: Chapter 3).

The Warring States also invoked religion to urge their soldiers into battle, but such attempts were not wholly successful. By this time the all-important ancestral cult emphasized the burial of an intact corpse and the avoidance of its decay in the tomb. Thus the state's claims on the bodies of its subjects were fiercely contested by those of the family, and no public entombment with funeral oration like that of classical Athens emerged to give ritual benediction to death in battle (Loraux 1986). Nevertheless, the Mohist canon notes the rulers' assertion that death in battle was the highest good, a military treatise emphasizes the importance of honouring and sustaining the families of those who died in combat, another insists that campaigns be preceded by divination in the ancestral temple to show the spirits' support for war, and a poem depicts the souls of slain warriors as eternal heroes amongst the dead (B. Watson 1963: 55; Sawyer 1993: 207, 223; Hawkes 1985: 116–17).

More significantly, religion underpinned the institution of warfare through the pervasive use of images of armies and combat in rituals. The most notable of these was the exorcism during the New Year season. Accounts of the imperial version of this rite relate that the exorcist and his assistants went through the rooms of the palace, stabbing into the corners with spears to expel demons and evil influences. This armed cleansing was completed by actual detachments from the capital army. Texts from the first century AD show that the common people likewise expelled demons from their houses with spears and other weapons. Similar military performances served to ward off drought and pestilence. Armed exorcists likewise preceded coffins into the tomb, and statues or carvings of them were left behind to guard against evil. Military action against demons also figured in the medical practice of the period, as we know both from literary accounts of folk practice and from medical writings discovered in a second-century BC tomb. Finally, tomb texts from the Han period invoke the spirit armies to defeat demonic forces that threaten the occupant of the grave. Stories regarding these battles against demons were incorporated into Warring States and early imperial mythology, and later texts suggest that these tales also had wide currency amongst the common people. Such routine appeals to combat for protection against spirit enemies could well have predisposed the Warring States peasant to view warfare as necessary and beneficial (M. Lewis 1990: Chapter 5; Harper 1982: 72–93; Seidel 1987: 678–714; M. Lewis forthcoming).

In summary, the writings of Warring States thinkers provide us with an extensive repertoire of procedures to create an army, induce men to fight and thereby make war possible. These factors include identifying the army with the whole of society so that to be outside the army was to be outside the human realm, applying rewards and punishments to elicit desired conduct, creating belief in a beneficent public order that must be defended, inculcating a sense of obligation to one's family and neighbours, teaching conventional standards of honour and shame, treating the enemy as criminal or demonic, fashioning a sense of *esprit de corps* through distinctive rituals and costumes, aestheticizing the military experience through art and music in order to create a dramatic spectacle for viewers and a feeling of unity for participants, forging a sense of belonging to a collective body through repeated drills, instilling personal devotion to fellow soldiers through common experience and to officers through shared suffering, offering status in the afterlife to those who died in combat, and asserting power over the realm of spirits through performances modelled on combat.

With the exception of the evidence from religious and medical practice, all the above ideas are preserved in writings by and for the literate elite. Assuming the centrality of warfare to the power of the state, or viewing it as a necessary evil in a corrupt world, these writers reflected on the means by which the peasantry could be induced to risk their lives for their rulers. These texts treat warfare as a matter of manipulation or control, in which institutions and policies existed solely to lead men, against their natural inclinations, to fight for the state. Our data provide no evidence of what interest the peasants themselves might have taken in warfare, but we can speculate about this on the basis of the fate of warfare and the Warring State as ideas in imperial China.

Although the institution of universal military service survived the unification of China, it soon proved unsuitable to the needs of a 'world empire' engaging in frontier wars with hostile nomads. Mass peasant levies gradually fell into disuse during the first century BC, were abolished early in the next century, and reappeared only in the twentieth century AD (Elvin 1973: Chapter 2; Lewis 1979). The dominant ideology of imperial China was resolutely anti-militarist, and the peasantry played no role in the armed forces. The army of the empire was regarded with fear and loathing by its subjects, and the prospect of actually becoming a soldier would have filled most people with horror. Nevertheless, the institution of a 'society organized for war' survived in important sectors of Chinese society, and the vision of combat and armies permeated all elements of popular culture. In these last few pages I will sketch the broad outlines of this curious phenomenon and reflect on what it suggests about the social bases of war.

The earliest non-official form of a society organized as an army was the late second-century AD millenarian rebellion of the 'Yellow Turbans'. Fighting in the name of the egalitarian vision of the 'Great Peace', this rebel movement was organized into military detachments under 36 generals. The leader of the movement bore the title 'Heavenly Prince General' and his two brothers were 'Prince General of the Earth' and 'Prince General of Men'. They, like the villages that became armed camps in response to the turmoil of the period, followed the military practice of swearing collective oaths to establish their laws. Such military organization and the swearing of oaths became a recurrent pattern in Chinese sectarian religious movements, many of them associated with the ideal of the 'Great Peace'. The last and greatest of these, the Taiping rebellion of the mid-nineteenth century, resulted in a greater loss of life than any war prior to the twentieth century.

As Wolfgang Bauer has noted, this military organization was not simply a function of rebellion but also an alternative to the social order presented by the dominant Confucian doctrine. This taught that the family was the fundamental unit of society and the foundation of social order. In such a family-based society, men lived under the fixed dominion of their fathers and were permanently tied to the confining circle of immediate kin. The army offered a radically alternative model: the primacy of youthful vigour over lingering decrepitude, a hierarchy based on energy and merit rather than seniority, and a society of peers linked in common service to a universalistic vision rather than the narrow concerns of the household (Bauer 1976: 118–19). This universalism gave a place even to women, who in sectarian movements and secret societies enjoyed the opportunity of participation and of rising through the ranks.

In addition to inspiring periodic sectarian movements, the spectre of armies and combat haunted the entire popular culture of late imperial China. The story-telling tradition and theatre, as well as the great classical and historical novels of the Ming (1368–1644), drew on a common body of plots and themes that were permeated by combat. Accounts of warlords and strategists from the Three Kingdoms period (AD 220–280) were the most widespread and influential of the historical fictions of combat, but tales of the military rise and fall of dynasties figured in many story cycles and historical novels. While accounts or performances differed depending on the social status of their authors and intended audiences, the more popular versions emphasized militarism, righteous revolts, universalistic associations based on loyalty among peers, and even the military skills of women (Hegel 1985: 112–42; Chang 1990: 181–95, especially 189). These narratives clearly suggest that the same features that underlay the military organization of millenarian religious groups were appreciated, in more conventional forms, by significant elements of the population.

In such a story appears the famous 'Peach Garden Oath', the classic version of the oath of sworn brotherhood which had a major impact on Chinese social relations. The historical incident on which this story was based was patterned on the collective oaths sworn by armies in the

Zhou and the Warring States. Through its prominence in fiction and theatre, however, this story spread throughout Chinese society and became a model for the oaths sworn by members of secret societies and other social networks outside those sanctioned by the state and the great lineages. And along with their oaths, these associations adopted the trappings of military organization.

The stories which form the classic novel *Water Margin* emerge from the same intellectual universe. People expelled from the conventional social order form a band of brothers through oaths of friendship. This band grows to become an army and ultimately a counter-empire based in an impenetrable lair which forms a microcosm of the world. The literati novel ultimately rejected the revolutionary implications of the narrative, but the profoundly subversive and totally military character of the fictional anti-world appears in most theatrical versions and lies just below the surface in the novel itself (Plaks 1987: Chapter 4–5).

Along with tales of societies constituted as armies developed a set of stories, novels and plays which were based on the military aspects of popular religion. The best known of these became the *Journey to the West*, wherein a group of animal spirits use their martial skills and magical powers to defend a pilgrim monk from the attacks of demonic adversaries. Another important group of tales deals with the demon-quelling warrior, Jiang Kui (Plaks 1987: Chapter 3; Eliasberg 1976). This use of military action to confront demonic forces also continued to figure in the religious rituals of the peasantry. These included the old military exorcism rituals, but they also employed the services of newer warrior deities, such as Er Lang, who had entered the pantheon over the centuries. (A film of militarized exorcism rites in contemporary Anhui province was screened at the Musée Guimet in Paris on 16 September 1992.) In these rites peasants availed themselves of divine martial powers to obtain protection, which the imperial state failed to give (Ahern 1981).

What, if anything, does this militarized imagination on the part of a demilitarized culture tell us about the social bases of war? It demonstrates that warfare is not simply imposed by rulers through the manipulation of ideology and the force of law. Chinese peasants found in the figure of the warrior an image of mastery, a figure of power and skill who could defend himself and others from hostile forces. In this way the fighters of late imperial Chinese literature were like the cowboys or detectives of modern popular culture. Tales of righteous rebel armies and outlaw bands offered an alternative world to the stifling reality of officials who never desisted and fathers who never died. Against the tyranny of kin and neighbour, the imaginary army offered a liberating vision of a community of like-minded peers in the service of a higher cause. As a refuge from the petty burdens of a tedious existence, the army promised immortality in a collective crowd-body committed to a vision of equity and 'Great Peace' (Canetti 1962). And for a poor peasant who spent his whole life toiling, bowing and scraping just to get by and get on, warfare offered the blessed opportunity to smash things up. Given the imagined capacity of warfare to offer all this — power, community, eternity and release — it may prove as hard to demilitarize society elsewhere as it did in China.

2. Payback and Ritual in War: New Guinea

Gilbert Lewis

Papua New Guinea became an independent nation in 1975, but it has not shaken off the stereotypes which hark back to a *Boy's Own Paper* world — savages, missionaries, cannibals and gold prospectors. Plumes and painted warriors are the images fostered in magazines and travel brochures. Such images reinforce the idea of a place where warfare is endemic and it becomes hard to keep these pictures from colouring other evidence and aspects of the people and the place. A great deal has been written on New Guinea, although few anthropologists have done their fieldwork in communities that are still actively taking part in warfare. In most of Papua New Guinea (PNG) fighting was stopped by colonial rule and the ethnographers came afterwards.

Pacification and Colonial Government

Spears and bows and arrows were no match for the power of guns. Even so, some commentators (A. Strathern 1984: 18–31) have wondered at the speed of pacification following first contacts, given the small number of patrol officers and police, and the size and diversity of the country. Quite large populations in the mountainous interior, the highlands, were not discovered or controlled until just before or just after the Second World War.

The experience of colonial government was thus quite short — in some places 30 years or less. Such societies contain people who had taken part in and remembered warfare before white outsiders came. Various groups in the highlands have gone back to fighting since independence: the central government has tried to contain and stop it, but without great success. In coastal and lowland areas, where colonial government had interrupted indigenous warfare for two generations or more, no people were left who could describe it from first-hand experience. The coastal and lowland areas of PNG have not shown a resurgence of warfare as have the highlands. Anthropologists may now work in areas where warfare takes place, but few do.

The aim of this book is to consider social forces and institutions that maintain war and make it difficult to stop. New Guinea has a reputation for endemic warfare although the fighting was suppressed in most parts of the country — by force initially (police, guns, gaol and punishments), followed up by political development and education. The new warfare now breaking out takes place in changed circumstances — with different possibilities of communication, mobilization, weaponry and organization, and also, of course, a different knowledge of the world outside the village. Most of the anthropological analysis of New Guinea warfare deals with a recollected past, not with warfare that the writer has observed. Exceptions to this include the work of Koch (1974) and Heider (1970) in the early 1960s in Dutch New Guinea (now Irian Jaya), and of Meggitt (1977) in the 1950s, 1960s and 1970s in the Enga region of the western highlands of Papua New Guinea. Warfare in colonial and post-colonial times may be different from that of pre-colonial times in scale, tactics, weapons and aims, becoming possibly more bloody and destructive (Blick 1988 — this is also suggested visually and unforgettably by the sequences of modern highlands warfare in the final film of the series made by B. Connolly and R. Anderson near Mount Hagen — *First Contact*, *Joe Leahy's Neighbours* and *Bitter Harvest*).

The Scale of Warfare in the Past

It is difficult to assess the realities of warfare in the past from recollections and oral testimony. Time lapse and forgetting take away certainty and detail, changed values distort descriptions of the tenor of daily life in the past. The island of Papua New Guinea contains about 700 different languages and much cultural diversity. The political units were often small — villages of 200–300 people might act independently. Clans of between 200 and 1000 people, although they might be linked with brother clans in larger alliances, would act as autonomous units and might fight with other clans within the same phratry (Meggitt 1977: Chapter 3). Whether village or clan, many of the political units that acted independently in warfare had populations of less than 1000.

Small numbers affect the character of war. This has to be kept in mind when reading about New Guinea warfare. It was not a matter of soldiers and armies but rather of hostile encounters between people from villages or clans not too distant or, indeed, neighbouring. The enemy were usually known or identifiable individually. The political units were also small. A village or a clan of 300 people would not have many more than 60 men able to fight, even if all who could took the field. Along the Sepik River and in some lowland areas (for example, the south Papuan coast), there were occasional villages of over 1000 people but very few such settlements. The alliances did not mobilize very large numbers.

Heider (1970: 78, 104–33) describes formal prearranged battles between two opposed alliances of the Dani in a densely settled area. The population of each alliance comprised 4000–5000 people and fielded about 400 men on each side. These formal battles are the largest engagements I have read about and had the character of ceremonial daytime encounters with challenges, display in self-decoration, rest periods. Battle might be abandoned for fear that rain would spoil valuable feather decorations (Heider 1970: 107). Although men were wounded and might die in such battles, it was not these formal battles but dawn raids, ambushes and, occasionally, treachery on visits, that led to serious loss of life. The formal warfare was less deadly than some surprise raids involving destructive indiscriminate killing. Conspicuous display, challenges, dances for the ghosts after battle deaths and the constraint of rules gave the formal battles something of the character of ritual or game: for instance, in the ritual warfare there were rules against firing volleys of arrows because they could not be dodged. Elements of ritual, ancestral invocation and threat display are often notable in formal warfare in New Guinea (Gardner and Heider 1969; O'Hanlon 1989). But nonetheless death from enemy attack was a prominent cause of death among the Dani. Raids, surprise and ambush were most to blame. Heider estimates that of all men's deaths, 28.5 per cent were the result of violent attack; for women the corresponding figure was 2.4 per cent (Heider 1970: 128). The Dani built high wooden watchtowers to look out for raiding parties slinking through the high grass along the no-man's-land between confederations of clans. The threat of raid and attack was a daily concern.

Reciprocity and the 'Payback' System; Participation in Offence and Defence; Lack of the State and Insecurity

The most obvious general point to note about endemic war in New Guinea is the insecurity, the absence in pre-colonial times of any central authority with power over the different political units or ability to control them. They were autonomous. There was no state. Political units organized themselves for defence and could count on no higher body than themselves for their security. They might make alliances on the basis of neighbourhood, common origin or ancestry, but there was no overarching force to guarantee the recognition of rights, observance of the rules or support. For order they relied on the principle of reciprocity, tit for tat, the right of equals to exchange gifts or blows, to match help or to retaliate (Brown 1979). Australian patrol officers called this the 'payback' system; in Tokpisin people said '*mipela bekim*' (we retaliate, we give it back in return). Colonial government and then independence brought some experience of centralized government, but the reality of it varied greatly in different regions of the country. In some places the people have never had much direct evidence of the power of the state to intervene, to guarantee their rights or to ensure their security; or reason to believe they could depend on it. Or that it would last. When talk of the state did not materialize in some strong local presence, people were more likely to go back to what they used to rely on. A more or less ephemeral experience of effective state control is at least part of the explanation for the resurgence and maintenance of highlands warfare.

The small scale of political units imposed a high ratio of military participation. All men were in principle expected to be ready to fight. The village or the clan is the defensive unit, but not necessarily the offensive unit. Threats to village or clan drew the whole political unit together, whereas revenge or retaliation might be decided on by individuals at a more personal level or in secret. The repercussions of such particular revenge could easily spread to involve others, for one faction's view of a just equivalent was not another's. Some were much closer than others to the dead, their grief and anger fiercer; in the reckoning of balance, the alternating disequilibrium could never settle at a point of balance that everyone would see as fair.

Only part of a political unit might be involved. There are many terms for violent encounters: fight, raid, ambush, battle, war, ritual combat, feud. They differ in scale, character and political organization. Are all to be included under the heading of war? And where should we place the witchcraft, sorcery and 'poison', which many Papua New Guineans saw as occult and deadly instruments useable in war? Sillitoe (1978) surveyed the different kinds of military engagement described in ethnographies of Papua New Guinea. He concluded they were basically divisible into battles with large numbers of warriors, and raids with small numbers using surprise in their approach. Battles and frequent raids occurred characteristically in regions of mixed forest and grassland, whereas infrequent destructive raids occurred in regions of swamp and dense rain forest. Battles are tactically impossible in a rain forest. Terrain and density of settlement affect the frequency of encounter and conflict. Tactics in offence and defence must take account of different patterns of settlement, for example nucleated villages, dispersed hamlets or scattered homesteads.

War Bound up with Ritual and the Question of Adaptive Value

The western and central highlands of New Guinea were densely settled areas, mainly of scattered homesteads. In these areas, it seemed plausible to link the maintenance of warfare with Malthusian competition, chronic land shortage or protein shortage (Meggitt 1977). Rappaport (1968) argued that an elaborate ritual cycle of alternating war and peace, which he found among the Maring, served to maintain their environment by adjusting the human population to the land through redistribution (rout in war) and adjusting the size of pig populations by periodic slaughter for the feasts which were part of the ritual cycle controlling times for war. He argued that their religious beliefs and ritual practices included war as an integral part of the ritual cycle: the whole complex emerged as an adaptive mechanism contributing to preservation of the overall environmental balance. But these explanations provoked counter-arguments and comparisons to show that the intensity and maintenance of war cannot be simply related to population density, land pressure or protein scarcity (MacArthur 1977; Sillitoe 1978; Knauft 1990). Other solutions to the pressures were reached without making war an integral part of the adaptation, for example by different systems of kinship and rules for land tenure and the reallocation of land (Kelly 1968). However, Rappaport's work brought more attention to the roles that taboos, ritual duties, cultural beliefs about ancestral ghosts and ties to land can have as stimulus or control in the organization and frequency of war. The relative importance of secular and religious forces in maintaining war seems to vary with region.

Political Rivalry in 'Big Men' Political Systems as a Factor Maintaining War

Sillitoe (1978) argues that traditional political leaders (usually called 'big men' in the literature) used war for their own political ends. They might have a scheming interest in maintaining war even though they disguised it. 'Big men' were not military commanders, but they could still influence events in war by taking the initiative in both aggressive and conciliatory action. They might try to manipulate situations of war to improve their own political standing: their positions were won in competition with rivals through their own efforts in persuading their clansmen to support their views about other clans, or whether to pay compensation for killings, or to start a further bout of war (Strathern 1971: 53–92). They needed to show outstanding energy and persuasive ability in influencing action and in organizing ceremonial exchange. War offered special opportunities for aspiring 'big men' to rise.

In many highland areas, especially those densely settled, clans formed alliances and defined certain other clans as traditional enemies. Clear identification of the alliances, the major and minor enemies, sustained the threat of war as a permanent possibility. The enmity endured as the clans endured, and their relationships of hostility contributed to their self-definition and solidarity. Rules about compensation for deaths and bounty payment for killings depended on the ally and enemy relationships. These relationships were further bound up with rules about preferred and forbidden marriage alliances. I shall not describe the detailed arrangements; they are intricate and exceedingly varied — some support the cliché, 'We marry our enemies', others its polar opposite — but the general point that would emerge strongly from the details is the way that war was part of a complex interlocked field of instituted relationships involving enmity and alliance, rules of marriage and expectations about the exchange of gifts, loyalty, blows, sorcery, poison and killing. The ties of women to their brothers and ties through women to maternal kin opened possible lines of communication, support and betrayal. These patterns had serious implications for the women in between (Strathern 1972), for relations between the sexes and the attitudes of sexual antagonism and distrust of wives for which some parts of Papua New Guinea are notorious (Herdt 1982; Whitehead 1986).

Male Cults and Ritual in the Maintenance of War

By contrast with much of the highlands, warfare in lowland New Guinea was less bound up with struggles over land and secular 'big man' politics. Rather, it had complex links with ritual, religious belief and cosmology, especially in parts of the south Papuan coast and the Sepik, where headhunting was practised. Swamp, dense rain forest, river or coast favoured infrequent destructive raiding, with long-distance and large-scale sorties (possible by canoe) against people of different ethnic groups. Some raids were related to the requirements for carrying out elaborate male cults concerned with male growth and the fertility and well-being of the people and their environment (Whitehead 1986; Knauft 1990). The raids involved selective predation and went outside the raiders' home territory. The people belonging to other groups feared those in the big river or coastal villages and kept well clear of their territory. The pattern of raiding and a frightening reputation suited expansion and gave locally dominant groups like the Marind Anim and the Iatmul relatively greater peace and security at home.

In the fringe areas of the highlands to the south and east, male cults assert men's control of mystical forces important for their growth and fertility. They separate the sexes early (at between five and seven years old boys are taken away from their mothers' control), and older men instil boys and young men with fear of female pollution and sorcery. Some groups hold long initiation ceremonies marked by male bonding, instituted homosexuality and the fostering of attitudes of sexual antagonism towards women. These have been interpreted as part of a masculinity complex linked to endemic warfare in very small groups. Each man had to be prepared for taking part. Each man might depend in some particular critical moment directly on the other man. Some of the autonomous fringe groups had populations of 250 or less (Herdt 1982; Barth 1990) from which to draw all their defenders.

The Insecurity of Daily Life: Expectation, Training and Incentive

I shall say something here about the Gnau among whom I worked, not because they were particularly warlike, but rather as an indication of what the insecurity was like, and of how experience and upbringing combined to prepare people for fighting and the risk of raids. I would suggest, first, that people accommodated to the insecurity and accepted risk as a normal part of daily life. If the risk seemed normal, then to be prepared, and well prepared, was normal and desirable. Secondly, I would note that men were led by their upbringing to expect to have to fight and were given strong incentives to be warlike. Actions and attitudes good for hunting helped in war — and vice versa — so the goals set for male achievement were locked together and the honours linked.

The Gnau belong to a small language group (about 1000 speakers) and live in three villages in the forested hills of the West Sepik Province. When I first visited them, 20 years after they had come fully under colonial government, many of the older men still carried a bow and arrow with them when they went to work in gardens. Some carried the barbed arrow used in fighting. They said they carried it for protection, just in case. There were still many signs of former times and the old insecurity and need for vigilance. The villages were placed at the top of a steep hill or by a cliff edge with narrow paths of approach. The incline just before entry was nearly always the steepest. Some of the approaches had had thick stands of bamboo planted to prevent access anywhere other than along the entry path. The fallen bamboo leaves made it difficult to come along it silently. Another path had had a deep pit set across it where it narrowed, spanned by a log, and this point, they said, had been guarded in the past.

The nucleated village was divided into hamlets, each with a conspicuous men's house. The doors at each end of a men's house were small so that a man would have to crouch to enter. All the men and youths of the hamlet slept inside. Wives slept nearby in separate small houses with their children. In the men's house, the beds nearest the door at each end were for senior men experienced in fighting. Bows and arrows, shields and spears were placed on a rack above each man's bed. The domestic houses were lined inside with firewood so that a spear could not be jabbed through the walls at someone sleeping inside.

Older men mimed how they used to work in their gardens, putting their bow and shield down near them, working away at the task (whatever it was), pausing frequently to look round and listen, moving the bow and shield forward each few yards to keep them constantly to hand. They described how the women would go to work in the sago groves with an escort of one or two armed men to protect them while they worked. Gardens were scattered in the forest surrounding the village. Families worked at separate plots. Nowadays families will sometimes sleep at the garden site where they have a house, but they did not do this in the past. They would have been too vulnerable. Calls and whistle-talk ricochet between families working near each other in garden land. The forest hides them but they can keep track of where the others are. Towards twilight people call out from the high point of the village to stragglers. I was particularly struck by overhearing a university-educated man from another part of the Sepik explain to his ten-year-old son why he had not answered him when he kept calling out for him one afternoon when his father was late back from a walk. The father said, 'I heard you and I answered you by a whistle. You should not have repeated your call. I did not answer you. Your repeated calls could draw attention to me, that I was somewhere outside, alone.

An enemy or sorcerer could hear and could hide in wait for me as I returned.'

They are still suspicious of strangers and the threat of sorcery. People can travel almost at will nowadays through other people's land. They do not like this. Along paths in garden land, they bury sharp barbed arrow points to pierce the feet of strangers or trespassers. I had to treat a number of people with infected foot wounds because of that. They may place signs of destructive magic to warn people off from paths, gardens and crops. But 'now we sit down like women,' they said, and *sanguma* sorcery men can go where they want. In the past, 'in the time of arrows', no strangers came on their land because they feared to. To have a reputation for being fierce and for killing gave one security at home: people would not want to risk coming on a raid. This was cited as a benefit of being warlike and successful in killing people from the neighbouring villages. Some of the older people had scarcely been outside their own village territory — perhaps a few brief visits to three or four other villages in their lives. Another reason stopping men from entering another village was danger from the spirits of the dead to men who had participated in killing someone from the other village.

When a man recalls what it was like as a child to be playing with his sister one morning in an almost deserted village and to see two black-painted raiders slink up over the cliff edge and run at him, we can imagine why people wanted a fearsome reputation. He ran shrieking, trying to drag his sister with him; she was pulled off his back and stabbed through with an arrow; he was being stabbed when men came running and the raiders fled. He was telling me the story of the scars on his belly.

The Praise of Killing: Hunting and Warfare Interlocked

When I first worked among the Gnau, quite often I saw older men wearing homicide ornaments at gatherings and rituals. Some came with a hornbill mounted with splayed wings, the great beak jutting out between their shoulder-blades, hung from their necks at their backs. A torque of pig tusks around their throat was more rare. Senior men conferred the right to wear these ornaments on their juniors according to their success in hunting and warfare. I saw the rituals carried out after successful pig kills and cassowary kills on a number of occasions. The rites for hunting were linked up with success in warfare. When asked to explain what would have given someone the right to wear homicide ornaments, they always stressed that he would have had to have killed many pigs and cassowaries as well as to have killed enemies. The rites were to bring success in killing; if a man went through the rites for homicide he would have success in hunting; if he went through rites for hunting, he would be successful in war. The two kinds of killing were spoken of as interlocked and alternating. They said they had modified some details of the hunting rites, leaving them out because they helped for war and that was forbidden now. But the hunting rites were still intended to make men fierce: they included bleeding, hot things, fierce nettles and teeth scrapings inserted in cuts along the wrist.

The ornaments and the marks of success in hunting are matters of great male pride. The skulls and bones of pigs and cassowaries are hung up for show at a man's day house where visitors come to see him, or they are hung up inside the men's house. Individual successes are publicly celebrated by hunting rituals for the man. A series of food taboos correspond to hunting success. Men eat together in groups and they are continually reminded by the food taboos of where they have got to in the series of taboos — a matter of prestige (G. Lewis 1980: 146–66). Male pride in killing was celebrated publicly through the rituals associated both with hunting and with warfare, and, by interlocking them, both were reinforced. These rules and rituals played a part in maintaining the values associated with war and they stimulated confidence.

There is another element to record. The line of pig skulls and cassowary breastbones also sometimes has some inconspicuous markers tied to it. They are short twigs — *nebigug* leaf petioles — tied to mark the people a man has killed by sorcery. They may be inconspicuous. But when a man dies, the drumming of the log-slit gongs (*garamut*) announces his death and beats out the list of his achievements. It goes on through the night and the dawn; it can be heard in surrounding villages. It beats out first the tally of people he killed in warfare, then the tally of people he killed by sorcery, then his cassowary kill count, then the pigs. Warfare was linked to killing enemies by sorcery as well as to success in hunting. They were all commemorated when they honoured the dead man and grieved for him.

Rituals for War: 'Wild Taro' and Trance

Hunting and warfare were linked by shared marks of prestige. Gnau rituals for puberty and male initiation were focused on growth and strength, health and success (G. Lewis 1980). The rituals had effects on male bonding through seclusion and shared initiation and a cult of male secrets and power. There were also rites specifically directed at preparing boys and men for war. The mother's brother had duties to strengthen and protect his sister's son. He carried out a ritual with a shield to make the boy steadfast and unafraid (crushing him down under the shield he presented to him). There were other rituals to strengthen men and make them fierce. Exceptional warriors were given a stone from the crop of the goura crested pigeon to eat so that this stone (*segedat*) would make the man restless and alert, tense as the quivering crest of the goura pigeon. A man with the *segedat* sees things faster, reacts more quickly, misses nothing, his eyes dart everywhere, he does not need to sleep as others do.

The main ritual or magical substance associated with war was a 'wild aroid' or 'wild taro'. Each clan has a special kind of wild aroid for war. The plants look like a giant rhubarb or taro plant. A man should shout out the name of his 'wild taro' when he killed (as I have heard with pig kills — the triumphant shout rings out when the arrow finds its mark and the animal falls). The wild aroids were planted in the village, but after colonial pacification the people destroyed them. So strong was their association with war that local people believed the 'wild taro' were banned by the white patrol officers. They planted them again, but hid them in concealed places in more distant gardens. The stem of the herb was chewed in rites following homicide and before fighting. They said the wild aroids burnt their mouths and throats, they would salivate, choke or vomit, and they might go into trance. Trance was the effect of the power in the plant. According to their theory, trance occurred when the man's thinking centre turned round and went into his back so he became oblivious of everything except the drive to kill. He would be ready to shoot at anything, would start at any movement, not recognize his kin. So in the rites after homicide in the men's house, the doors were shut and the homicide was held down by a shield.

They thought that the wild aroid made some men go into trance in warfare itself. I once saw a man begin to go into trance as tempers grew very hot in a large confrontation over land: he was quickly held by men on his own side. In the past, they said, men who went into trance were dangerous, they were reckless, unreliable, deaf to calls or appeals. They struck fear into enemies. But those who were really admired as warriors were not men who went into trance. Great fighters were, in their view, those who were wild and daring, able to stand heedless of danger but constantly alert, fast and fierce. But there was much ambivalence in attitudes to the fierce trance of men in warfare.

In a detailed study, Harrison (1993) has presented an interesting analysis of attitudes to war in the large Sepik River village of Avatip. He focuses on war magic in which they also sought to put fighters into trance to strike terror in the enemy. He notes that trance was supposed to make men capable of extreme violence. It did so by making them blind and deaf to persons; it anaesthetized their perceptions; it enabled them to deny the enemy's humanity. Trance removed normal social controls on killing and was a device to relieve the fighter from accountability for killing. In contrast to Hobbes's view of war as the original natural state of man, which has to be controlled by the imposition of political order, the outlook of Avatip people, Harrison argues, takes sociality for granted and the problem for them is to make men capable of war. For war, they must deny normal values of sociability and recognition of other people's humanity. They have to be prepared for violence. The purpose of the men's cult and the special ritual and magical measures taken in Avatip is to sustain proper levels of warlikeness. The aim is to overcome humane scruples and construct aggressiveness through ritual. War requires opposite attitudes and values to those on which normal social life depends. Violence is given value only in certain contexts. The men's cult was the means to create this violent male ethos. Working nearby among the Iatmul, but two generations earlier, Bateson (1958: 161) noted something of the same ambivalence to violence — the violent man may be admired but he is also a liability. The needs of war and the needs of peace are not the same. Real leaders must be resolute, not reckless.

Two-edged Issues

The ambivalence towards violence is one example of a double-edged element in many of the issues here. The double-edged element in many of their institutions contributes to making war difficult to end.

Societies in Papua New Guinea have lacked an overarching central power, the state. Modern highlanders might argue that warfare is not necessarily to be looked on as some sort of breakdown of social order, but on the contrary is sometimes a good response, the right one for the problem. The real question is the nature of the problem. War, in their view, may be the appropriate way to solve the issue.

Without the state, people were free but insecure. Without the state, reciprocity widely served as the principle of order between autonomous equals. It could be a friendly or a hostile reciprocity — an exchange of gifts or blows. There was danger in excess and danger in restraint. But it is difficult to agree about fair exchange — 'payback'. Among the Gnau, I could not sort out any clear balance sheet on deaths and revenges from what the interested parties said. Who counted for which death? The closer the kin, the hotter the anger. And the action of a hothead could destroy some fragile balance.

The scale of war and politics in New Guinea is sometimes so small as to make it a matter of almost individual action. The tiny scale demanded preparedness from everyone. They were all at risk of involvement. If all men had the right to, and were ready to, use violence, that added to security and it added to insecurity. Fierceness and violence were needed and feared.

Excess and Restraint

People adapt and adjust to insecurity. But if they do it too well, they may come to accept too much insecurity as normal. The socialization and rituals for boys and men were meant to prepare them psychologically and physically to face external threats, to create dependable loyalty, and to give them skill in war and defence. The institution of rewards for aggressive behaviour or killing and the opportunities for aggressive display are evident in the ethnographies. Ritual display and game-like formal warfare may, like animal threat displays, allow the release of excitement or the expression of threat; the displays may aim to deter, but get out of hand and go beyond it.

Finally, in this chapter, I need to identify a major omission. I have almost totally neglected the women's view of warfare. I have written about factors that maintain war only from the men's perspective. I have not tried to describe the part women are said to play in support or betrayal, nor the ways that endemic warfare affected their position, the attitudes towards them, sexual antagonism or dependencies. Women, I should emphasize, are often placed very differently in terms of their loyalties to the clan or village groups in which they were born and those where they end up as wives and mothers. The weights of their attachments shift with time and change as their families grow up and their children marry. Many of the men's priorities, politics and aggressive ambitions would not be shared by their wives.

3. Communal Violence in Northern Ghana: Unaccepted Warfare[1]

Susan Drucker-Brown

Over the past two decades conflicts in northern Ghana have increased in frequency and destructiveness. Most are unreported in the national media and educated Ghanaians are loth to think of them as 'warfare', a label that might imply fragility in the nation state.

Some 37 different languages are spoken in northern Ghana by people with radically different political traditions. Former kingdoms (Mamprusi, Dagomba, Nanumba, Gonja, Wala) coexist with peoples who have no traditions of centralized government, some of whom were encompassed as subjects within those kingdoms, while others were autonomous at the turn of the century when the British invaded. Conflicts occur both within and between ethnic groups. Intra-ethnic conflicts are usually 'dynastic' disputes between related groups claiming kingship or chiefship; inter-ethnic conflicts occur most frequently between members of chiefly groups confronted by peoples they regard as subjects. To Ghana's southern population, all northerners seem remote and are represented in the media as moving slowly, with government aid, out of poverty and backwardness. Few southerners have visited the north. Fewer speak a northern language. Northerners, by contrast, are more likely to have visited the south for work or education; many of them know one or more southern languages. Reversing the southern stereotypes, northerners regard the south as uncivilized, inhabited by bloodthirsty savages. The word for people who come from the south also refers to musket-bearing foot soldiers. When used for southerners it is best translated as 'barbarians'.

This pejorative association between weapons and southerners indicates the value northerners place on peace. Throughout most of the region, blood shed on the earth is an offence against the earth as divinity, which should be repaired by sacrifice. In ordinary disputes onlookers will intervene to prevent individuals from fighting physically. Homicide is not only abhorrent but regarded as polluting and mystically dangerous to the perpetrator. But the north was not always peaceful. In the former Mamprusi kingdom, during the mid-1960s, the oldest people could remember warfare — the invading European soldiers with their African mercenaries and terrifying weapons, the warlords who resisted the Europeans and raided markets and villages for slaves (Goody and Drucker-Brown 1993). The Mamprusi were both victims and raiders, but wars for slaves were regarded as shameful. By contrast, fighting among young men of different villages or village sections, arising from the abduction of women, was described by old men more as sport than as lamentable violence, though the fights led occasionally to deaths and the destruction of property. Compensation could be paid for homicide and sacrifices performed to re-establish peace and cleanse the earth.

Battles among rival princes in competition for kingship or chiefship were regarded as an ongoing form of warfare. However, by the 1960s secret sorcery and poison had replaced set battles. In the past, horses were trained for warfare, princes fought with swords and foot soldiers used muskets, spears, bows and poisoned arrows. Gunpowder was still being made locally in the 1960s, but the Mamprusi bemoaned the fact that there were fewer horses about. Bows and arrows, muskets or, increasingly, rifles, were used for hunting but, as hunting declined in importance, weapons seemed to be mainly of ritual significance. At a man's funeral, his bow and quiver full of arrows are exhibited and, during the final phase of the funeral, muskets are fired to praise the deceased and show the grief of the in-laws. If the deceased was an important person, gunfire is used to break down part of a compound wall in order to perform the burial. If the funeral is of political significance, the number of guns is still carefully counted and calculations made to compare the strength of rival candidates to chiefly office. For the most part, however, the notion of people keeping weapons for fighting would have seemed bizarre in 1965. Reports of armed confrontations seemed part of a distant past. I was told that previously trade and travel were dangerous, but that the white men had brought peace and that this was a good thing. It should be emphasized, however, that this was before 1966, when the first military *coup d'état* displaced the first independent Ghanaian government headed by

Kwame Nkrumah and the oscillation between civilian and military regimes began.

Generalizations about the outbreak of violence over the last 20 years can be reached only through a detailed understanding of each conflict. This requires reference to details of history and production, chiefship rivalries, economic competition, demographic and ecological changes, and relations with central government.

The B'moba–Konkomba Conflict

On my return to Ghana in 1985, I found a much impoverished country and a grimly obtrusive military presence. I saw missiles being driven into the presidential palace and airports full of soldiers bristling with weapons. In the north, the central market of the regional capital at Tamale had been recently burnt by the army, I was told in retaliation against a protest by market-stall owners. Chaddian refugees were begging in the streets. In 1986 the atmosphere had changed again, at least superficially. An IMF loan had put commodities back into circulation. The Tamale market, rebuilt, was larger and more flourishing than ever, though the roads were much worse than they had been in the 1960s. More transport was available but it was more dangerous. Radio communication and telephone posts had been set up, but the postal service was almost non-existent. The army presence was more subtle, but nonetheless unmistakable. Nalerigu, where I lived from 1963 to 1965, had grown from a village of 4500 people to a town of 12,500. Some running water was available where formerly women had to walk a mile to the river, and grinding mills had been introduced. But greater poverty was unmistakable in the daily diet.

I had come to attend the Mamprusi king's funeral, an event that elicits fierce competition among princes. The capital was tense. An army officer came to offer the elders in charge of the funeral a detachment of soldiers for 'peacekeeping'. The elders tried hard to persuade the army to stay away. In the face of their polite discouragement the officer finally remarked that the army would come even if not invited because: 'My men also like to eat meat at funerals.' However, it was not among Mamprusi that violence occurred. While the funeral was in progress, fightng broke out 15 miles east of the capital among B'moba and Konkomba people, minority ethnic groups within the former kingdom.

Inadvertently, because I had not thought I was visiting a war zone when I asked to accompany the local doctor, I found myself on a mission to 'identify the dead', inspecting the smoking remains of a string of burnt-out villages. Refugees were fleeing towards the king's village, some packed tight into lorries with leafy branches tied to the drivers' cabins and sides of the vehicles. The leaves, I was told, marked the lorries as 'neutral'. Clearly, this was no unique event. The four dead we found were elderly men, all killed by gunshot — 'hunting rifles', the doctor said. Two men had their throats pierced by arrows and their left hands removed. 'Traditional Konkomba custom,' said the police. The dead were B'moba. The combination of modern and ancient techniques in their treatment parallels the mixed sources of this conflict.

Although they occupy an area the Mamprusi king nominally controls, the chiefships held by Konkomba and B'moba peoples are a British artefact. District Commissioner Amherst (1947) recorded misgivings about a newly created chiefship at Bimbago. The village, jointly occupied by the B'moba and Konkomba, had one chief for each group. He was concerned that neither chief was able to collect taxes and that Konkomba 'malcontents' were leaving the village. He wrote of 'the election of a headman', but Amherst depended on the Mamprusi king to install these 'headmen'. In the Mamprusi kingdom, chiefship or headmanship was never an elective office, but depended on genealogy combined with competition (Drucker-Brown 1975). The fact that 'malcontents' were moving away indicates resistance. Though it has now become more difficult to move, this is still a typical form of opposition to an unwanted chief.

Amherst's disquiet was well founded. The fighting between B'moba and Konkomba in 1984, 1986 and 1989 is said to have been triggered by a single quarrel in the Bimbago market. A quarrel began between a woman selling mangoes and a young man buying them. In some versions the woman is B'moba and the man Konkomba, in others the identities are reversed. All agree, however, that the mango seller insulted her client and a fight between young men followed. But descriptions on either side are more concerned with the specific acts of destruction and retaliation that involved whole settlements. This is the story of the conflict as told by the Konkomba chief of Bimbago in July 1991:

> There is no fight now. Only hunger. But in the past, the B'moba people started it. They claimed that the chief in Bimbago was a B'moba and everyone knows the Konkomba own Bimbago. One Konkomba chief made a B'moba man *ba'anaba* [chief of the soothsayers]. It was just a name. The Konkomba chief gave the B'moba land for farming, for building, and they were strangers

and came to the market. The Konkomba own the market and collect the market tax. But they were united [the Konkomba and the Bmoba] and it [the market] was for everyone. Then the *ba'anaba* started calling himself the chief of Bimbago. The B'moba stopped coming to greet the Konkomba chief. They were controlling the market. The Konkomba were being asked to pay the B'moba for tickets. That is one reason for the conflict. The money is not much, 10 cedis, 100 cedis, 200 cedis, it is for the council. But it means, if they collect it, they own the village.

The B'moba came in March 1986 and burnt the whole town. People ran away. In 1989 they came again and we fought against them with local guns and cutlasses. They were using machine-guns, and guns from the police and army. It ended with 46 policemen staying at the school and we were feeding them. Every two months they were changed and they stayed for over a year. Gradually the number of police was reduced to four. Now the only B'moba in Bimbago are women, two of them married to Mossi men. The four teachers and the nurse were B'moba but they ran away (leaving the settlement with no trained nurse or teachers).

The B'moba version, July 1991:

The fighting was all due to chieftaincy disputes. There was a Konkomba chief and a B'moba chief living on opposite sides of the river. The Konkomba said the whole area was his. The B'moba said it was all his, and he wouldn't pay allegiance to the Konkomba. Fighting began when the Konkomba attacked and the B'moba ran to Nakpanduri [a neighbouring B'moba village]. The Konkomba burnt down houses in the territory of the Nakpanduri chief. The question then was how could we retaliate? The B'moba gathered from many villages, and joined together and surrounded the Konkomba villages. The government had to intervene.

Neither of these chiefs spoke English. The B'moba chief was much younger than the Konkomba one. During the interview, he was surrounded by a group of young literate men, including an articulate and angry school teacher. They said they were not prepared to accept the situation as it stood. The Mamprusi king had intervened after the third episode of fighting and deposed the B'moba chief, leaving only a Konkomba chief in the village. The motive of revenge is emphasized by both groups, along with a desire for the chiefship as a symbol of dominance. In local political rhetoric 'ownership' means the right to deference, and it is the honouring of a chief through greeting and the giving of 'gifts' that manifests deference.

In each episode between ten and 20 men were killed and fighting seemed to continue till numbers were roughly equal. In the three clashes between Konkomba and B'moba only men were killed. In the last episode (1989), a 'tank' (an armoured personnel carrier) was sent from Tamale. Leaders of both communities were detained and imprisoned in couples, one each from opposing sides, and then asked to sign an agreement that they would be personally responsible for seeing that no further fighting occurred in their communities. In some villages, jointly inhabited by Konkomba and B'moba, peace has been re-established. There has been a ritual 'burying of the blood' with sacrifice to the earth and an oath of reconciliation among former enemies. But in Bimbago there has been no reconciliation.

The market has been a focus of tension in many disputes in the north, but those involving Konkomba people are particularly bitter. Konkomba are well-known as hard-working farmers who shun high population densities, preferring to move on to thinly populated land and make new farms rather than to stay in settlements where, as population increases, the markets become larger and more profitable. Konkomba farmers, however, use markets for the sale of their yams and livestock. B'moba people have had more exposure to town life and more education. They tend to remain near markets and are more likely to control them. However, as the Konkomba chief indicated, the market tax is small and its symbolic significance is what matters. For both groups the deference commanded by a chief is clearly signficant in itself.

The Range of Conflicts in Northern Ghana

Following is a summary of the four conflicts which occurred during the last six months of 1991. They give some idea of the range, scale and motives of communal violence.

The Dintigi-Kandiga Health Centre

In June there was a single clash among Kassena people from the settlements at Kandiga and Dintigi. The two settlements are contiguous, lineages of a single clan with a common ancestor. The local health centre was named 'The Kandiga Health Centre'. Young men of Dintigi removed the sign and replaced it with one that read 'Dintigi Health

Centre'. The placard was changed several times and in the fighting, which eventually occurred, cutlasses, bows and poisoned arrows, and modern rifles were used. Some 18 young men were killed. Grain stores were burnt and animals slaughtered. The authorities detained elders from both sides for several months and no fighting has yet recurred.

In considering this dispute it is necessary to refer briefly to the structure of the national administrative bureaucracy in Ghana. The country is divided into regions, and regions subdivided into districts. Each district has its own officers for public services, and since 1989 each district has collected taxes and disbursed funds from central government. Since independence the number of districts has risen steadily. The rapid rise, from 110 to 200 between 1989 and 1992, was officially part of the military government's devolution of power to the 'grass roots' in a system of elected district assemblies. The government decreed, however, that new districts might be created only if three criteria were satisfied. There must be (1) a given level of population, (2) an appropriate infrastructure, and (3) a peaceful population.

The rational calculation behind the fighting relates to the second of these. The Kassena people (a linguistic/ethnic group that includes both Kandiga and Dintigi) had been agitating for a new district. A particular building contractor is said to have paid young men from one sector to change the name of the health centre. This followed an earlier episode in which the market was moved from its customary outdoor site in one section to a market building in the rival section. The section hoped that with both market and health centre it could lay the greater claim to 'an appropriate infrastructure' and become the site of the district headquarters. Clearly this dispute is, at least in part, about who will have access to resources coming from central government. As such, it is the simplest case of a much more common phenomenon.

In the B'moba–Konkomba conflict described above, district organization was also a factor. During the period of CPP government (1958–66), partly as a result of a chieftaincy dispute in Bawku (see below), this territory of the former Mamprusi kingdom became an independent district with a B'moba district commissioner and a district headquarters in the largest market town. When Nkrumah was deposed the Nakpanduri District was reintegrated into the South Mamprusi District. District administration is an important source of employment and other revenues for local people, particularly literates. Some observers suggested that B'moba politicians might encourage conflict in order to reclaim their former district. Others suggested that the Mamprusi might encourage the fighting between the B'moba and Konkomba to prevent any re-emergence of the former district. Whatever the truth, the shadow of the former district, and the more tangible presence of the former district commissioner, still alive and locally resident, were both regarded as elements fuelling the conflict.

Dagomba Chieftaincy Dispute

Early in the year fighting took place in the Dagomba towns of Karaga and Gushiego. Some 40 men were killed, including police sent to stop the fighting. The supporters of a deposed chief of Gushiego clashed with the followers of a newly installed replacement. Gushiego people fled north into the East Mamprusi District. A village chief there reported in June that 50 refugee families had arrived in his village. It was the end of the dry season and food reserves were at their lowest. He petitioned the regional government for aid. Central government declared the district 'a famine area' and some food was sent. From June through August soldiers enforced a curfew at Gushiego. Early in September fighting occurred in the much smaller Dagomba village of Vogu (population 6000), where another new chief was to be received. The fighting is said to have been well prepared and both factions well armed. Russian sub-machine-guns were mentioned. This, like the Karaga–Gushiego conflict to which it is linked, was an episode in a dispute among rival factions claiming the Dagomba kingship, whose origins predate the colonial period (Staniland 1975; Ladouceur 1979).

Today's succession disputes, however, have been crucially affected by new ritual procedures. In all the former kingdoms of northern Ghana, the installation procedure depends on activating links between his ancestors and the initiated chief. There is no customary means of deposing a chief, thus decrees that depose chiefs are locally regarded as illegitimate. In the Gushiego case, moreover, the deposed chief is a well-known politician, a former chief inspector of police and member of a former national military junta. He was deposed by the present government as part of a process that awarded the contested Dagomba kingship to a rival faction. As in the Kusase–Mamprusi dispute described below, rival families are allied to long-standing national political factions. Thus all aspiring chiefs and their supporters are polarized and any change in a high-ranking chiefship, or the kingship, polarizes the entire system because all Dagomba chiefships are linked in promotional series and through procedures that require certain chiefs to select and install others.

The Nawuri–Gonja Dispute

Between early April and late June Kpandae and surrounding smaller settlements were burnt and looted in two separate clashes between the Gonja and Nawuri peoples. The Gonja are a numerically and politically dominant minority in a large, ethnically heterogeneous area. They claim descent from invading founders of their kingdom. The Nawuri claim autochthonous origins near the market town of Kpandae. The Gonja regard the Nawuri as subjects. To this political division a religious one must be added. Catholic missionaries have converted a Nawuri flock, while the Gonja, though not universally Muslim, regard themselves as closer to Islam than to Christianity. Fighting in Kpandae is said, in Nawuri circles, to have been triggered by a Gonja chief's threat to sell the land on which the Roman Catholic mission is built. In Gonja circles, the Catholic mission priests are supposed to have aroused the Nawuri with sermons on liberation theology.

There were two successive episodes of fighting in 1991. During the second episode, when warning must have been given, individual compounds and even sections of compounds occupied by people who were neither Gonja nor Nawuri were marked with green leaves. Care was taken by the combatants to avoid damaging these. The victims are said to have been nine Nawuri and one Basare man, and ten Gonja. A woman was killed by a stray bullet. The death of a Basare man at the hands of the Gonja was regarded as particularly dangerous, since the neighbouring Konkomba people, who are closely related to the Basare, might have been drawn into the conflict. The Konkomba are much feared as a result of the massacre of the Nanumba in a series of Konkomba–Nanumba conflicts in 1980/1. Ironically, the Gonja called on the neighbouring Nanumba chief of Bimbila to negotiate with the Konkomba and present apologies for the death of the Basare youth. Another particularly provocative death was that of a Nawuri youth living some distance away in the Gonja town of Salaga. Between the two episodes of fighting at Kpandae, he was attacked by a mob and beheaded. This attack was not accompanied by fighting and seems to have been part of a Gonja attempt to terrorize the Nawuri.

A third, and the most serious, episode in this conflict occurred in June 1992.[2] According to one version, a group of some 30 well-armed Gonja, accompanying a new chief to Kpandae, seized produce and animals from the market and killed a Konkomba storekeeper. In an alternative version the violence began when Nawuri people destroyed a petrol station and killed the Gonja owner. Both versions agree that an alliance of Konkomba, Nchmuru, Basare and Nawuri peoples fought the Gonja arrivals who sought refuge with the police. The Gonja were massacred, their bodies mutilated, the chief beheaded and the Gonja weapons seized. Two more lorries filled with Gonja men *en route* to Kpandae were ambushed. Major Gonja towns were looted and burnt and thousands of Gonja refugees fled to Tamale. The army, including special reserves from other parts of the country, were called north. Kpandae, Kpembe and Salaga were occupied by troops. At Kpembe, the father of the regional commissioner of the Northern Region was village chief. Twelve members of his family were killed and the family house was totally destroyed.

The Tamale Riot

In early July 1991, on the night of the annual Dagomba 'Fire Festival', soldiers returning from enforcing the curfew at Kpandae drove into a celebrating crowd in Tamale. At least eight civilians, including a child, were killed and several wounded. One soldier was beaten to death by the crowd. The following day the regional commissioner's home was sacked and government offices destroyed. Beer bars in the market and surrounding streets were burned. Military men from Accra appeared on national television visiting the wounded in hospital and urging local leaders, stony-faced Dagomba chiefs and Muslims, to keep calm. The burial of the dead took placc in the presence of an all-male company of mourners. The absence of women indicated that violence was feared. A curfew was imposed on Tamale for the next five days.

The riot, though it clearly differs from the previous cases, was not unrelated. The soldiers were returning from policing the Gonja–Nawuri area, only recently the scene of fighting. This conflict had made everyone aware that the regional commissioner was the son of an important Gonja chief. The riot involved the local Dagomba population of Tamale, already angered by an incident that had occurred earlier in the year. In March a Dagomba trader had been shot and killed by police when he had refused to stop at a checkpoint. Local people pointed out that neither the police, nor any of the Gonja mob that had killed the Nawuri youth at Salaga, had been punished. It was said that the regional commissioner could never discipline the Gonja because his father was a chief, nor could he proceed against the police or members of the army because he needed their support against local people (i.e. the Dagomba). In addition to anti-government feeling generated by the arbitrary behaviour of the police and army, the government decision in

1986 to overturn a House of Chiefs recommendation awarding the disputed Dagomba kingship to one faction of Dagomba royals aroused widespread resentment. It was said that even the faction that had benefited from that decision would prefer a Dagomba to a Gonja regional commissioner. In the aftermath of the 1992 conflict, which had led to the deaths of his relatives, the Gonja regional commissioner resigned and was replaced by a Dagomba.

Finally, the riot elicited stereotypical northern fears of southerners and somewhat more realistic fears of northerners among the southern residents of Tamale. The driver of the army van, a southern soldier, was said to have been drunk and to have publicly announced before he drove his van into the crowd that he 'wanted blood'. This story, which is consistent with the northern stereotype of bloodthirsty southerners, was circulated not only in the north, but among northerners in Accra. Southerners living in Tamale feared they might be the butt of retaliation, but this did not occur.

The Mamprusi–Kusase Conflict

The relationship between Mamprusi and Kusase peoples shows how different types of conflict can become superimposed. It paralleled the B'moba–Konkomba conflict in that local chiefship, at least in part, was a British artefact. Like the Gonja–Nawuri conflict, a chiefly group was opposed to traditionally chiefless peoples. Finally, it paralleled the Dagomba disputes in that the ever-widening polarization of local groups was tied to national events. The conflict centred on the chiefship of Bawku, an ancient market town. In the bush outside Bawku is the tomb of the founder of the Mamprusi kingdom and the site of a former Mamprusi capital. At the turn of the century there were six Mamprusi chiefships in the region of Bawku, all six of which were installed by the Mamprusi king. They resided in an area occupied mainly by 'Kusase' people, who themselves had no traditions of chiefship. During the mid-1930s, when the colonial government set up an administrative centre in Bawku, it was decided that only the chief of Bawku should be installed by the Mamprusi king and he should install the other five Mamprusi chiefs, as well as a newly created set of Kusase chiefs. A new hierarchy was thus established.

The population of Bawku town itself has always been composed mainly of immigrants connected with trade. Today the Mamprusi chiefly family speak Hausa rather than Mampruli. In contrast, the farming area surrounding Bawku is densely populated by groups that are now labelled 'Kusase'. That label was originally an outsider's name for a population of interrelated clans and lineages with diverse origins, living in dispersed settlements. This population has gradually acquired a common 'Kusase' identity and conflict with the Mamprusi has been a major context for this new-found solidarity.

In the Ghanaian parliamentary elections of 1957, Kofi Busia, an Ashanti aristocrat, recruited the few Mamprusi politicians into his United Party. They had the support of the Mamprusi king. Kwame Nkrumah and the CPP were supported by many northern groups without any chiefly traditions or aristocracies, though, as we shall see, the correspondence was by no means absolute. When the United Party won the 1957 elections in Bawku, known supporters of the CPP were victimized. Some CPP members lost government employment, others were given jobs in distant places, while the Mamprusi who had supported Busia remained in or near Bawku and advanced their careers. In 1958, while tension was growing among the educated young men who supported opposing political parties, the Mamprusi chief of Bawku died. Rival princes lobbied the Mamprusi king and a successor was chosen. There were several eligible candidates, especially two Nkrumah supporters and one Busia supporter. The king chose the Busia supporter. The first phase of the installation ceremony was performed in the capital. A second phase remained to be performed in Bawku. However, before the newly installed chief arrived home from his installation, he was effectively replaced. A Kusase member of the dead chief's household was persuaded to undergo an *ad hoc* ceremony carried out by the dead chief's court, using regalia still in their possession. The instigators of this coup were an MP nephew of the dead chief and a Kusase district officer, both members of the CPP. Their Kusase chief was officially recognized as the Bawku chief. Thus, the first Kusase chief of Bawku was installed by an alliance of disappointed Mamprusi royals and a Kusase district officer.

A realignment of northern politicians took place after the 1958 elections. The Mamprusi left the United Party to join the CPP and the Mamprusi king was threatened with deposition if he failed to support the CPP government (Kelly 1974) and in 1960 the Northern Region, which had till then encompassed both Bawku and the Mamprusi capital, was divided into two separate regions. There were other administrative reasons for the division, but when the Kusase district officer of Bawku became the first regional commissioner of the newly created 'Upper Region' and the Mamprusi king was forbidden to install Mamprusi chiefs in that new region, the Mamprusi king and court concluded that the administrative change was a punishment for the king's choice of an 'opposition' chief for Bawku. This division of Upper and Northern regions was to fuel many conflicts in an area where princes tracing Mamprusi origins contested chiefships of mainly ritual importance. When these groups were displaced by new chiefs from neighbouring kin groups, conflicts among related heirs to chiefship gave way to confrontations between groups that defined themselves in terms of different origins. Paradoxically, such groups were often far more similar culturally to one another than the groups tracing Mamprusi origin were to the Mamprusi.

In Bawku, as the Kusase began to rally in support of the new Kusase chief, events on the national scene served to unify the formerly divided Mamprusi. In 1960, at Kulungungu, near Ghana's northern border and close to Bawku, there was a nearly successful attempt to assassinate Nkrumah. Among others implicated in that attempt was the family of the Mamprusi chief unsuccessfully installed in 1958 by the Mamprusi king. In the aftermath of the assassination attempt, all Mamprusi chiefs in the Bawku area were jailed or fled into exile, and were replaced by new Kusase chiefs. Kusase chiefs who had been installed by any Mamprusi chief of Bawku were also jailed and replaced. They were kept in detention until 1966. With the fall of the Nkrumah government in that year, Mamprusi chiefs were released from detention, or returned from Togo. Their Kusase replacements were deposed and they were reinstated. The assassination attempt and its aftermath thus extended the conflict over chiefship throughout the Bawku area, and it merged with the already widespread rivalry over jobs among educated young Kusase and Mamprusi men.

Since 1966 every change of national government has been accompanied by fighting in the Bawku area. Sometimes only parts of Bawku are destroyed, at other times surrounding villages are burnt. Each episode is followed by a period of army occupation. The coup that deposed the Busia government reinstated the Kusase chiefs and deposed the Mamprusi. Several Mamprusi chiefs have taken cases to the courts. Others still hope to acquire chiefship by force. Weapons employed are increasingly destructive, and with each episode more young men are drawn into preparations for revenge.

Discussion

In all these conflicts, excepting the Dintigi–Kandiga dispute, chiefship is a common thread. When former subjects 'rebel' against chiefs, they are not dismissing the institution of chiefship but demanding chiefship for themselves. Chiefship has increasingly acquired a similar function to the name of the health centre, both as an assertion of identity and a promise of access to government resources. In terms of resources this is no more than a promise. Chiefs receive a small stipend but do not dispose of government funds. However, chiefs are a channel of communication to central government. Particularly in the absence of a civilian government, chiefs and their courts are a focus for popular lobbying at village level and they also serve to disseminate government decisions. But this still does not explain why people are prepared to fight for such offices. It is true that chiefships can provide opportunities for individual profit-making. Conflicts between chiefly and chiefless peoples are often fuelled by chiefly demands for 'gifts'. But most of those involved in chiefship battles have no individual aspirations to office. They fight for chiefship because chiefs continue to be, above all, a symbol of corporate identity. In dynastic disputes among rival claimants to a single chiefship, victory ensures a kind of immortality for one section of a family. Those who claim their own chiefs in defiance of others clearly see chiefship as a symbol of their equality with chiefly groups. Finally, control of chiefship also implies control of land, as we shall see below.

The deposition of chiefs, which has fuelled many conflicts, immediately links the local and national levels of government. It is a step taken by national government 'to bring the locals into line'. Northern politicians lobby their colleagues in Accra for what they see as their local advantage, but the deposing of chiefs that results as national governments change, creates major problems. Traditions of northern kingship and chiefship are unanimous that installation vests an individual with office for life. There are no traditional procedures, other than execution, for deposing a chief or king. This is not true of chiefship elsewhere in Ghana. Southern politicians making decisions about the replacement of chiefs in the north and coming from traditions in which chiefs may be deposed, have perhaps been unaware that each new chief would become a permanent office holder and sets of alternative chiefs be created, waiting and no doubt working for a return to some previous regime. This has happened in Dagomba, Bawku and elsewhere in the north, where there are often two chiefs for each village chiefship.

The changing demographic balance between ethnic groups within districts, as well as the general population increases, clearly plays a part in these conflicts. In much of northern Ghana chiefs allocate land. Land has been plentiful in much of the north and in such places it is an offence to deny land to someone who wishes to use it to feed himself. Hence, despite the growing population pressure, chiefs continue to welcome strangers in a tradition of seeking always to increase their followings. When, as in the Nanumba–Konkomba case (also the Gonja–Nawuri case and possibly the B'moba–Konkomba case), the population of recent immigrants comes to outnumber its hosts, the former 'strangers' may decide that they are in fact the proper 'owners of the land'. 'Ownership of land' is increasingly regarded as synonymous with chiefship, though there are other sources of rights over land. The transformation of local ecology similarly exacerbates pressure on land. In the Nanumba–Konkomba conflict, the building of the Volta Dam flooded farmland and may have blocked the path of Konkomba expansion southward.

The problems resulting from cumulative migration and population increase, which change the balance between ethnic groups sharing a single territory, may also be related to the fact that conflicts seem to occur more frequently near international borders. In certain conflicts the *de facto* dual citizenship of members of an incoming group is said to be significant to that group's military strategy. Thus, the Kusase of the Bawku area and the Konkomba of the Mamprusi area are bitterly resentful that their enemies have been able to cross the border and find support among relatives on the other side. Borders are, of course, patrolled by the army and smuggling is an important feature of the local economy in all market towns near the border. This may lead to 'arrangements' between soldiers or police and local people, both for the acquisition of weapons and for free passage across the border.

It is often said by local people that weapons used in these conflicts come from the army or the police. This is a possible source, but arms are frequently bought from private dealers who freely sell unregistered weapons. A full-grown cow is worth between £80 and £100. A sophisticated rifle (AK47) is worth roughly the same. In persistent conflicts the sense of grievance felt by opposed groups is translated into the regular collection of funds for the purchase of weapons and the payment of lawyers where chieftaincy disputes are in the courts. Ordinary farmers contribute, usually cattle.

Conclusion

It is paradoxical that, though local conflicts are linked to national events and party politics, they have the effect of drawing people's interest away from the national to the local scene. People who have suffered the burning and looting of their villages are more directly concerned with the results and future of these conflicts than they are with the state of the nation. But northern preoccupation with parochial politics is by no means a rejection of national identity. Most northerners, literate or not, identify themselves as Ghanaian citizens. Particularly for younger people, identification with the Ghanaian state is a source of pride and a symbol of modernity.

As yet these localized conflicts have not coalesced to form a single movement, though diverse groups demanding or defying chiefs show signs of coming together, as in the alliance of Konkomba-Nawuri-Basare-Nchmuru, which fought the Gonja in 1992. This, however, is not a direct expression of anger against the government, as happened in the Tamale riot of 1991, or the subsequent post-election riot in Tamale in 1992. Some observers consider that populist government rhetoric has deliberately encouraged 'anti-chief' conflicts in order to deflect criticism from itself. However, most of these conflicts, as the concern with chiefship indicates, are about the recognition of particular local (ethnic or family) claims at the regional or national level on a footing of equality with others apparently more favoured.

Though many of the conflicts in northern Ghana can be shown to emerge from conflicts which predate the nation state, they are now inescapably embedded in Ghanaian nation-state politics, and an economy which is part of the industrial capitalist world (cf. Tambiah 1986 on Sri Lanka). Gellner's observation (1983) that 'language and education are part of the technology of industrialization' is surely important here. Literate members of groups in conflict are often the most militant just as they are the most vulnerable in their dependence on national government which, in northern Ghana, is practically the sole employer of their skills. Clearly, the acquisition of Ghanaian identity, which they most strongly express, does not erase other allegiances.

The Ghanaian material suggests that political structures are far from discrete. People have overlapping identities and are subject to pressures that emerge from the diverse political structures in which they are enmeshed. The recognition of competing local or family groups at a national level may be the prize but, in the violence that results, revenge may become more important and local antagonisms create new group boundaries. The mixture of weapons and techniques of warfare used in northern Ghana reflects the diversity of political process. The fact that the as yet uncapitalized 'subsistence agriculture' of the north is producing the cash to buy sophisticated weapons, which are increasingly used to carry on these conflicts, is an economic aspect of the same superimposition of political forms. Ancient antagonisms and new aspirations lead to new hostilities, which are inextricably tied to the nation state even as they may threaten its integrity.

Notes

1. This chapter is based on recent research funded by the Economic and Social Research Council during 1990–93, as well as on earlier work in 1963–5 and brief visits between 1985 and 1987.
2. I am grateful to Alicia Fentiman, who witnessed the conflict at Kpandae, and Bernard Beirlich, who was resident in Tamale during the riot, for allowing me to use their fieldnotes.

4. Nationalist Imaginings of War in Cyprus[1]

Yiannis Papadakis

During the last 40 years Cyprus has provided the setting for a number of violent confrontations related to attempts at national self-determination and external interventions. This chapter seeks to explore the darker sides of Greek Cypriot and Turkish Cypriot societies by concentrating on the language, symbols and forms of their respective nationalisms, the aim being to show how violence and war are embedded in nationalist rhetoric and imagery. As Anderson points out in his seminal study on nationalism, it can inspire both love and hatred. However, Anderson's view (1991: 142) regarding the insignificance of 'elements of hatred in expressions of national feeling' I find contestable, as feelings of glory and love cannot be separated from those of hatred and animosity, at least in the specific model of nationalism under discussion here. Moreover, as I try to show, the social mechanisms that lead to a diachronic transference and exaggeration of an ethnic group's suffering, along with a tendency to ignore or belittle another's, can lead to polarized situations where compromise or dialogue has few chances of success.

Symbols of Nationalism

If, as Anderson (1991: Chapter 2) argues, the idea of the nation has replaced those of the dynastic realm and the religious community as the prime source of moral authority in the modern world, its imagery and moral prescriptions deserve close examination. In Anderson's own view, notions of sacrifice and war are central to the nationalist imagination. Kapferer's study of nationalisms as diverse as those of the Sinhalese and the Australians (1988: 135) confirms Anderson's insight of the centrality of images of war and sacrifice, to the point where it leads to 'the glorification of war'. These observations are clearly borne out in the context of the primary symbols of the nation state — the flag, the national anthem, the parade — that the two sides in Cyprus utilize.

As far as the flags and the anthems go, both Greek Cypriots and Turkish Cypriots have in the recent past mostly relied on those of their respective 'motherlands'.[2] The version of the Greek flag most commonly used has nine stripes, which symbolize the nine syllables of the slogan 'Freedom or Death' ('*Eleftheria i Thanatos*'). The notion of sacrifice and blood is embedded in the design of the Turkish flag. According to the common account provided in Turkish school textbooks, it symbolizes the reflection of the crescent and a star in a pool of a dying fighter's blood. The Greek *Hymn to Liberty* talks of Liberty springing from the 'holy bones of the Greeks', recognizable from 'the playing of that terrible keen blade'. The Turkish *Independence March* asks Liberty not to frown upon the Turkish 'heroic race' so as to deserve the blood that was poured in its name.

On both sides, military parades mark the most significant national commemorations. Such events provide occasions for the glorification — to the point of explicit worship — of diverse types of arms in mass yearly public rituals. Such parades are set in a heroic atmosphere of flags and marching tunes praising the glory of the nation and the courage of its fighters, sometimes being accompanied by live commentary providing information on the use and function of the weapons, along with a stream of praise for the nation and descriptions of the period and events being commemorated by the parade. That the most important national commemorations are marked by military parades in itself points to key values at the heart of nationalism. If nationalist rhetoric constantly alludes to a 'deep horizontal comradeship' (Anderson 1991: 7), what better way than a parade to express this, where 'the nation' marches as one, in the same step and rhythm with the same uniform and in the same direction? The 'historic atmosphere' that surrounds such occasions, through their allusions to a specific historical event, through the historical background provided in the lyrics of the marches, or through the commentary accompanying the parade, brings us once more to Anderson's (1991: 30) view of the 'nation as a solid community moving steadily up (or down) history'.

Nationalist Narratives of the Past

One reason why national symbols are intimately linked with preoccupations with war and violence is to be found in the nationalist conceptions of the past. Such conceptions often present war as the very stuff from which history is made, thus coming to 'naturalize' it as one of the most significant elements of human existence, a notion with important implications for the persistence of war. This section explores some aspects of the nationalist presentation of the past in the form of narratives of national struggle. Though this is not a universal model of history, it is how history is popularly conceptualized in many parts of the world, the Balkans being just one obvious example. Drawing from the two Museums of National Struggle, situated on either side of the divided capital, Nicosia, as well as from the ways in which history is commonly presented in school textbooks, I try to indicate how such nationalist constructions of the past lead to undue emphasis on past armed confrontations, create dehumanized images of other ethnic groups/nations and conjure images of 'historic enemies', which lead to associated notions of revenge or retribution.

Detailed descriptions of these museums are provided elsewhere (Papadakis 1994); here I confine myself to the issues that concern us. These two museums present the past of Cyprus in radically opposed ways by subsuming the island's history into the histories of the respective motherlands with which the groups in question either desired to unite or whose protection or help they were seeking. The Greek Cypriot museum was built in the 1960s to commemorate the five-year-long Greek Cypriot armed anti-colonial struggle which, despite its aim of *enosis* (union with Greece), culminated in independence in 1960. The Turkish Cypriot museum was built in 1978 and concentrates on the inter-ethnic violence, which followed independence until the 1974 coup launched by extremist Greek Cypriots with the help of the Greek junta, which led to the military intervention of Turkey. The division of the island followed by population transfers allowed Turkish Cypriots to set up their own state in the north, aided by the presence of the Turkish army.

Despite their striking differences these museums draw from a common idiom (as their shared name suggests) in presenting the past as a narrative of national struggle. Narrative discourse, being an account of reality governed by continuity rather than discontinuity, requires a social centre by which to locate events with respect to one another and charge these with moral significance. Thus narrativity is taken to be intimately related 'to the impulse to moralize reality, that is to identify it with the source of any morality we can imagine' (White 1990: 14). If, as Anderson argues, the idea of the nation is nowadays the major source of morality, it is this that features as the major actor and moral centre of these narratives. The continuity of the narrative requires the existence of an actor who appears as continuous or the same. The idea of the nation comes to assume this role and, as a result, it is presented as coterminous with history and in this sense is 'eternal'. Then the emphasis on 'struggles', by which is meant armed struggles, as the contents of the two museums (which concentrate exclusively on fights, fighters and guns) suggest, presents a view of the past dominated by violent confrontations. Finally, the notion of 'struggle' presupposes the existence of enemies.

The major problem with such presentations of history is that they are not viewed as one possible way of presenting the past among others (for example, social history and art history), but provide the paradigmatic form of what is in Cyprus commonly understood by the concept of history itself. Thus, both Greek Cypriot and Turkish Cypriot school textbooks closely follow the historical form the two museums provide. Each side's history is constructed by posing its own nation as the moral centre, according to which events can be judged as tragedies or victories. While it would be more correct to name such narratives as the histories of Greeks or Turks in Cyprus, they are in fact presented as histories of Cyprus. Treating the history of a geographical space or of the corresponding state(s) as equivalent to those of nations amounts to a conflation of the two concepts, the state and the nation, in a coterminous whole, namely the nation state. But by treating the nation and the state as coterminous, the presence, experience or desires of other ethnic groups within the state is effectively ignored, or they appear only from the standpoint of one's own ethnic group/nation as obstacles, internal enemies or residues from foreign occupations. The treatment of other ethnic groups as foreign to the body of the nation state permits, in cases of inter-ethnic confrontations or general anomie, attempts to 'cleanse' the national body (that is the state) from such elements. During more peaceful periods this may take the 'milder' or more symbolic forms of linguistic purification movements, folklore studies that seek to establish what is authentically 'ours' and purge culture of other elements, along with the neglect or plain demolition of monuments treated as 'remains left over by foreign conquerors'.

The narrative aspect of such histories and their emphasis on struggles also give rise to perceptions of other groups as 'historic enemies' who also appear as continuous or the 'same'. If one can draw pride and joy from the nation's past accomplishments, one is similarly entitled to feelings of animosity or pain for what such 'enemies' have committed against the nation in the past. Since the concept of enemy implies some kind of contact, it is mostly specific neighbouring polities thought to have committed past injustices against one's nation state that are placed in this role, whether there are current disputes or not.[3] This in itself presents the danger that any dispute could be seen as a manifestation of eternal confrontations making the process of negotiation more difficult ('how much more goodwill can we show?') or leading to an escalation by being treated as 'the last straw'. The problem is augmented if one considers the mismatch between the maps of states with those of ethnic groups, which means that neighbouring polities will commonly include minorities taken ethnically to 'belong' to the adjacent 'nation state'. The modern history of Cyprus provides a good example of the potential for internal conflict, as well as for the external 'protective' or 'enabling' military interventions — which in 1974 both Greece and Turkey felt obliged to undertake — that can arise in such a situation. Doubtless, latent notions of 'historic claims'[4] on the island, along with the idea that there is 'something wrong with the present borders between the two countries' (Millas 1991: 30), provided internal legitimation for both these interventions.

The dates the two ethnic groups in Cyprus chose as the 'beginnings' of what each regards as 'The History of Cyprus' reinforce the belief that the very presence of the other ethnic group on the island is somehow problematic. Greek Cypriot history begins with the settlement of the first Mycenaean colonizers on the island during the fourteenth century BC. Portraying Cyprus as 'a Greek island from the dawn of history' gives rise to the view that Turkish Cypriots are simply 'remains from the Turkish conquest in 1573' and thus not truly part of the indigenous population. The relevant start for Turkish Cypriots is the coming of Cyprus under Ottoman rule in 1573. As the Ottoman rule lasted until 1878, Turkish Cypriots in turn argue that Cyprus has been Turkish for three-quarters of its history. What the two versions of the past are in fact saying is that 'historically Cyprus has been ours'. Thus, a meaningful dialogue between the two ethnic groups is difficult to achieve, for both feel that the very presence of the other is in some way illegitimate.

Hatred and Love

If this is the past to which the prime symbols of nationalism refer, their preoccupation with war, glory and suffering is hardly surprising. Notions of heroism and liberation presuppose those of the enemy and oppression. For it is only in the fight against an enemy that heroes can emerge and only against a background of oppression that liberation can be meaningful. In this context, love of the nation implies identification with its past sufferings, respect for the heroes and their actions but also hatred for the enemies. It is this that makes Anderson's view regarding the absence of feelings of hatred in expressions of national feeling problematic. While it is true that national occasions concentrate on the glory of the struggle and the self-sacrificing love that the nation inspires, if the basis for the construction of the past is a narrative of national struggles, such occasions will inevitably give rise to associated feelings of pain and hatred against the 'enemies'.

Celebrating the victory of 'our nation' in the course of such commemorations also signifies the celebration of the defeat of the 'enemy/oppressor'. Thus, the celebration of Greek independence by Greek Cypriots was simultaneously a celebration of the defeat of the 'Turkish oppressors', and the celebration of the independence of Cyprus one of the defeat of the British and their 'Turkish Cypriot collaborators'. Similarly, the Turkish Cypriot celebration of the independence of Turkey was directed against the 'Greeks and their Western allies', while the celebration of Turkish Cypriot independence in 1974 is presented as a victory over the Greek Cypriots.

Heroes and Villains

If nationalism does resemble religion in providing answers to vital questions of values, life and death, it is the national heroes who are immortalized as saints in its pantheons, the Museums of National Struggle. Sacrificing oneself for the nation becomes the road to symbolic immortality, with the role of the hero presented as the noblest for youth to follow. If the walls of primary school classrooms are to be decorated with important personalities, more often than not it is photographs or paintings of heroes that predominate in Greek Cypriot classrooms and, as far as I could tell, those of the Turkish Cypriots. Educational curricula not only present the fighter/hero as a role model but also put forward the idea that 'ours is a nation of heroes' — heroism and bravery emerging as unique national characteristics.[5] If the notion of a powerful enemy is required to establish the heroism of 'our people', this can only be due to their superior numbers or armaments — an idea that both museums are at pains to stress by providing displays of the powerful modern guns 'they' used next to the ones of poor quality 'we' had to fight with.

While one's own national struggles are always presented as noble and glorious, the enemy is presented as barbaric. The cover of the primary school textbook on Greek history (Aktypis et al. 1990) used by Greek Cypriots presents an evocative illustration of such themes. A group of Greek fighters holding a flag (under which the national anthem is written) and waving swords is heroically presented against a background of 'Turkish oppressors' beheading a group of Greek priests. Turkish Cypriots have devoted a whole museum, the Museum of Barbarism, to the actions of Greek Cypriots, while the latter's Museum of National Struggle similarly presents Turkish Cypriots as 'slaughterers' who 'commit barbaric acts'. As Millas (1991: 25) also notes in the context of his review of Greek and Turkish school history textbooks (which are the ones used by the two ethnic groups in Cyprus) 'superiority is only relative and requires that the other party be discredited'. He subsequently finds both sets of textbooks equally guilty of striving to establish the 'superiority of our nation' (1991: 24–5) and of 'belittling the other' (1991: 25–7).

The most striking illustration of these notions is found — hardly surprisingly — in the army arithmetic lessons, where I, for example, learnt that 'one Greek equals ten Turks [mainlanders]'. Turkish Cypriot informants spoke of having received the reverse lessons of nationalist mathematics during their military service. Significantly, Loizos (1981: 57) points out that it was on the basis of such calculations that the extremist-nationalist members of the EOKA B organization, which carried out the coup leading to the Turkish military offensive, dismissed the Turkish threat. This lends credit to Stoessinger's (1985: 207–8) observations on the role of misperceptions of the other's capabilities and distorted views of an adversary's character in precipitating war. Such distortions and misperceptions, however, are part of the structural logic of such nationalist constructions of the past. Thus nationalists on both sides often deride those who talk of the need for realism, negotiations and mutual compromises as 'cowards' who have not learnt the 'lessons of history which prove that our nation has always emerged victorious against insurmountable odds'.

The Past as a Source of Post-Independence Political Legitimation

The impact of such nationalist narratives as elements in the socialization of individuals in a world that places the utmost importance upon the concept of the nation — as the misnomer 'United Nations' for example suggests — deserves serious consideration. But the question of why history is presented in a form that places an exaggerated emphasis on conflict, violence and sacrifice as a source of national pride has not yet been answered. Here, I present an explanation for this in relation to the post-independence political struggles in Cyprus, leaving it to those better acquainted with other geographical areas to evaluate its relevance elsewhere.

Accomplishments in the arts and sciences are not completely ignored, but if they do feature at all in school histories they do so only as short chapters or appendices, while the bulk of the syllabus deals with political history. Surely in Cyprus the influence of Greek and Turkish histories, which Greek Cypriots and Turkish Cypriots took as their models, is significant. One of the generic problems with such constructions of the past is that they are mostly concerned with 'history from above' (Millas 1991: 29), which, by focusing on the evolution of empires and states and concentrating mostly on 'wars and spectacles' and political events at the top level, ignores the grassroots. The perception of war as a supposed structural necessity has a long tradition in Western thought and is by no means confined to these societies (cf. Pick 1993). Periods of peaceful coexistence and elements of cooperation and syncretism, which a social history could reveal, are completely ignored (also cf. Loizos 1981: 642). Considering the two museums within the socio-political context of the periods in which they were set up can reveal some of the underlying motives that guided the forms they subsequently took.

The Greek Cypriot armed anti-colonial struggle for Union with Greece, led by the fighters' organization known as EOKA, ended in 1960 with independence being granted to the island as a bi-communal republic. Left-wing Greek Cypriots from the communist party, AKEL, who had initially expressed their reservations about both the form and timing of the struggle, were eventually excluded and became subject to considerable violence from EOKA (Purcell 1969: 270–1). The employment of Turkish Cypriots as auxiliary policemen also led to considerable inter-ethnic violence between Greek Cypriots and Turkish Cypriots opposed to the former's plans for Union with Greece (Pollis 1979). Turkish Cypriots thus began to voice their own demands, initially for the return of Cyprus to Turkey and later for the partition of the island. Independence found the two ethnic groups having to cooperate, with bitter memories of the previous inter-ethnic conflict and preferring to pursue their own goals.

The decision to set up the Greek Cypriot Museum of National Struggle was taken in 1961. Greek Cypriots were still keen on Union and this is reflected in the historical narrative of the Greek Cypriot museum, which presented independence as unjust and the struggle as not yet finished. After independence it was mostly the ex-EOKA fighters who came to monopolize positions of power in the Greek Cypriot administration (Markides 1974: 69; Attalides 1979: 55; Loizos 1974: 122), though by the early 1970s this was not necessarily the case. Their claims to positions of power were mainly challenged from two sources: members of the educated elite who had not necessarily fought but had the best educational qualifications for governmental positions (Loizos 1974: 123–4) and the communist party, AKEL. Throughout the pre-1974 period, AKEL did not receive any high posts in government (Markides 1974: 44). Its stance on the issue of Union was ambivalent. It traditionally had the best links and relations with Turkish Cypriots and more than any other political force followed a policy of reconciliation with Turkish Cypriots (Markides 1974: 62).

The form and content of the Museum of National Struggle can thus be seen as an attempt to consolidate political authority and silence alternative voices in the context of internal Greek Cypriot politics. By glorifying the struggle and the fighters (whether dead or alive), by ignoring intra-ethnic killings and by presenting the struggle as pure and heroic, it provided a reply to those who challenged the claims of ex-fighters to positions of political authority. Portraying the enemy as powerful and brutal served to underline both the contribution of the fighters in liberating the people and the extent of their efforts and sufferings. The implicit argument was that, since it was they who bore the costs and who were responsible for bringing liberation, they were the ones who deserved to be in control of the new state. The communists who had not participated in the armed struggle should have no claim to power in the newborn republic. Moreover, abandoning the ideal of Union would amount to an absolute betrayal of all those who died in its pursuit, and seeking the cooperation of the Turkish Cypriots could, in this light, be regarded as an equally treacherous act of cooperation with the enemy.

These comments, and the ones following concerning the Turkish Cypriot museum, are not meant to imply that atrocities were not committed or that oppression did not exist, and they are not trying to belittle the idealism of many of those who took part in such struggles, in which the over-representation of youth is easily established simply by looking at the photographs of the dead in the two museums (Markides 1974).[6] Rather, what concerns me here is to indicate the political and economic uses to which these struggles were subsequently put, as well as to explain the forces that influenced the forms that both the museums and the dominant historiographical models assumed.

The Turkish Cypriot museum was constructed in 1978 in commemoration of the Turkish Cypriot struggle. It stresses the events of inter-ethnic violence from 1963 to 1967 and concentrates on the suffering of Turkish Cypriots at the hands of Greek Cypriots and the resistance struggle of the fighters of TMT, the Turkish Cypriot equivalent of EOKA, with photographs of those who died prominently displayed. The 1974 'happy peace operation', as the Turkish offensive is called, is presented as having delivered them from the Greek Cypriot yoke and allowed them to set up their own state. In contrast to the Greek Cypriot museum, whose narrative did not attain closure as it implied that the struggle should continue, the Turkish Cypriot narrative distinctly finishes with 1974 standing for the final act in any self-determination struggle, independence itself.

A number of the points previously discussed regarding the relation of the form and content of the Greek Cypriot museum to the internal political contests of the period of its construction apply also to its Turkish Cypriot equivalent. Admittedly, due to the brevity of my research with Turkish Cypriots and a lack of alternative sources, I have less evidence of the post-1974 Turkish Cypriot leadership's links with ex-fighters than I have of the Greek Cypriots'. However, the links of the UBP, the party which emerged as the major political force among Turkish Cypriots and which virtually monopolized the leadership, with ex-TMT fighters were noted by both its opponents and its followers (also cf. Coufoudakis 1983: 119).

The claims of former fighters to economic and political resources have, I should point out, been a consistent pattern of post-1960 politics in Cyprus, where clubs for EOKA and TMT fighters still proliferate and act as focuses for the presentation of their demands. Apart from legitimating the post-1974 leadership, the museum's narrative serves as a means of silencing alternative voices among the Turkish Cypriots, notably from the left which is more in favour of reaching a federal settlement (Soysal 1992: 42–3), by presenting this as a betrayal of the sacrifices of those who fought for Turkish Cypriot independence. To those who oppose the claims by the Turkish Cypriot left that Turkish Cypriots and Greek Cypriots can coexist, this narrative readily presents the argument that history clearly 'proves' that Greek and Turkish Cypriots cannot coexist because the past is solely one of conflict. It also exposes the left to accusations that it seeks once more to throw the Turkish Cypriots into the arms of the barbaric Greek Cypriots.

Both the Greek Cypriot and the Turkish Cypriot left have tried to put forth an alternative vision of history that stresses — and sometimes romanticizes — peaceful struggles (such as common strikes) and cooperation at the grassroots. These ideas dominate the parties' and their followers' political discourse. That these positions come closer to what could be regarded as 'social history' rather than 'nationalist history' is not accidental. Nor are they apolitical as, within the context of both sides' domestic politics, they attempt to stress the positive aspects of their own contributions, which they feel are excluded by the dominant paradigms of history. Since cooperation at the grassroots in an institutionalized setting took place mostly through institutions of the left, such as trade unions, and since the left of both ethnic groups came under attack from the nationalists, these institutions came to reinforce feelings of mutual solidarity and of opposition against the nationalists of both sides. Still, their attempts to revise history present an effort to claim a space for the representation of grassroots experience completely denied by the dominant visions of the past, an experience the academic works of both insiders (for example, Kitromilides 1988) and outsiders (Sant Cassia 1986) are increasingly highlighting.

The Diachronic Transference of Pain

One outcome of constructing a state's past as a narrative of national struggle from the perspective of one ethnic group is that memories of pain at the hands of former oppressors are diachronically transferred to younger generations, even if they refer to events that took place hundreds of years ago. At the same time the sufferings or experiences of other ethnic groups are ignored. The subsequent impression is one of 'it is only we who suffered', which gives rise to notions of 'historic wrongs' and to latent feelings of revenge. This places any current conflict in a series of sometimes seemingly 'eternal confrontations' and leads to an accounting of pain such that each side may feel that its 'historical sufferings' justify a position of extreme 'self-righteousness'. This can seriously undermine the chances of success of any process of negotiation.

This mutual 'self-righteousness' was one of the most striking features of my research with the two ethnic groups in Cyprus, during which I was able to observe how even events within living memory are given radically different significance. The 1974 division of the island

and the Greek Cypriot desire to reunite it have led to a revision of the history presented in the Greek Cypriot museum. Greek Cypriot official rhetoric now stresses a past of 'peaceful coexistence with Turkish Cypriots'; this is to justify the vision of a united Cyprus and to counter Turkish Cypriot claims that the past proves that the two ethnic groups cannot live together.

This has led to silence about inter-ethnic conflict during 1963–7. Since younger people are hardly aware of these events, they are passing into the realm of social amnesia. In contrast, the Turkish Cypriot leadership talks of little else, and the Turkish Cypriot textbook on the history of Cyprus (Serter 1990) is basically a list of whom Greek Cypriots killed during this period and how. The scale of these events in the social memory of each ethnic group is readily apparent from the terms commonly used to describe them: if they talk about them at all, Greek Cypriots dismiss them as mere 'incidents' ('*episodhia*'), while Turkish Cypriots often refer to them as 'war' ('*savash*') and associate them with a period of fear, fighting, deprivation and involuntary dislocations. The public commemorations that inform social memory (Connerton 1989) on the Turkish Cypriot side place significant weight on these events, while Greek Cypriot social memory focuses on the catastrophic consequences of the 1974 division of the island brought about by the coup, and on the 'barbaric Turkish invasion', as the Turkish offensive has come to be called (Papadakis 1993). But for Turkish Cypriots this was the '1974 peace operation' which delivered them from Greek Cypriot oppression and, as the term itself implies, they showed little understanding of its impact on Greek Cypriots. Yet even the Turks and Turkish Cypriots revert to the use of war symbolism if they wish to glorify these events in commemorative documentaries. Similarly, the 'peace operation' turns into 'war for peace' in the title, with an imagery of background mementoes into which the photographs of Turkish soldiers who fought in Cyprus, or who are serving there, are inserted.

Moreover, while on the Greek Cypriot side mention of the 1963–7 period is avoided or treated as treacherous, the same applies to talk of coexistence with Greek Cypriots on the Turkish Cypriot side. While it is true that memories of pain and oppression may be exaggerated — or selectively presented — for various reasons previously discussed, denying the other group's sufferings as mere propaganda only reinforces feelings of 'self-righteousness' and hinders the emergence of a mutual understanding that could lead to a successful resolution of the conflict.

Conclusion

In this chapter I have not presumed to present a complete explanation of the war and of armed confrontations in Cyprus. Rather, I have tried to indicate some of the factors that promote violence as a legitimate means of solving disputes, that augment the scale of violent confrontations, or perpetuate them, as well as those that prevent conflicts being resolved before reaching the stage of armed confrontation. For such processes to be successful an understanding of each other's experience, sufferings and fears is required, and I have attempted to identify some of the obstacles to reaching this.

The role of such portrayals of the past in contributing to the eruption, escalation or perpetuation of armed conflict by presenting war as the motor of history and as the most significant source of national values with associated ideas of 'historic enemies and debts', might thus deserve serious attention in contexts beyond Cyprus, as the multiple sources of ethno-nationalist conflict nowadays remind us. If, as Hobsbawm (1992) argues, in times of social disorientation ethno-nationalism becomes a potent force, then such views of the past, which he calls 'retrospective mythologies' (1992: 3), have an ominous potential.

Notes

1. This chapter is the result of fieldwork with Greek Cypriots and Turkish Cypriots carried out in various periods between 1990 and 1993. For their financial support I would like to thank the Leventis Foundation, Smuts Memorial Fund, Richards Fund and Wyse Fund. For their comments I am most grateful to Professor P. Kitromilides, Dr P. Loizos, Dr M. Kunt and R. Fsadni. All responsibility for the views expressed above is my own.
2. Both states, the Greek Cypriot Republic of Cyprus and the internationally unrecognized Turkish Cypriot State, also have their own flags, which they use alongside the flags of Greece and Turkey respectively. Turkish Cypriots also have their own national anthem, though it is rarely used.
3. If the independence struggle were an anti-colonial one, then the former colonial authority would also be presented as one of the enemies, although the geographical distance between the two states would tend to reduce the significance of such enemies.
4. In Greece these may take the form of posting Cyprus at the 'edges of Hellenism' ('*i Akritiki Meghalonisos*') while in Turkey Cyprus may be treated as part of the 'historic Ottoman lands' and is sometimes referred to as the 'child-land' ('*yavru vatan*'), which implies the need for care and protection by the 'motherland' ('*anavatan*').
5. See Frangoudaki (1979: 115) for the betrayal of the Greeks as a nation of heroes in Greek education and Mardin (1978: 241) for analagous concepts in Turkish education and society. The broader dissemination of these ideas

in Greek culture is presented in Koliopoulos's (1987: 293–325) analysis of the concept of the 'brave handsome young man' ('*pallikari*').

6. Markides (1974) shows that this was statistically the case for the Greek Cypriots. I have been unable to obtain similar statistics for Turkish Cypriots.

5. Families in Conflict: Pervasive Violence in Northern Ireland

Dominic Murray

One of the more remarkable aspects of the conflict in Northern Ireland is that it stubbornly refuses to follow the development observable in most other ethnic wars. The normal pattern elsewhere has tended to be that eventually there occurs a sudden or significant rise in violence which is of such proportions that some form of compromise and at least partial accommodation is forced upon the antagonists. In comparative terms this has not happened in Northern Ireland, and it has been argued that it is because the levels of violence have remained more or less manageable or acceptable that the conflict has proved so enduring. It is therefore interesting to speculate as to why the conflict in the region is unique in this important respect. One possible explanation is that people there have become habituated not only to images of violence, as have people in Britain, but to the violence itself. Since now there are whole families who have had no experience of a social context without violence, habituation to it should come as no surprise. Also it should be stated that the warring factions have much more in common than is ever publicly admitted (a common distrust of the British government for example). In addition, there are many shared activities, in the form of sports and pastimes for example, which are indicative of some form of common culture. Finally, even for participants, it is not always easy to distinguish friend from foe.

At the outset, it is essential to state that the conflict in the province is not principally a religious one. It is true that it has been presented as such throughout the world but this is both misleading and not very productive. Within it there are of course religious elements which include a general suspicion on the part of Protestants in the north of the island with regard to what they perceive to be the religious hegemony of the Roman Catholic Church in the Republic of Ireland. However, the essence of strife is not religious but rather a result of differing, and often mutually exclusive, political and cultural aspirations. Nevertheless, it has now come to the stage that the labels of 'Catholic' and 'Protestant' are commonly employed by the media to compartmentalize the conflict neatly and conveniently. An example might best demonstrate how misleading such labels can be. Some years ago, the following advertisement appeared in the *Belfast Telegraph*: 'Wanted reliable cook-general ... Protestant. (Christian preferred)'. It might appear that the advertisement would make more sense if the words Protestant and Christian were transposed. To do so, however, would be to rob the advertisement of its intended meaning. In fact, the word Protestant here has little to do with religion but rather is descriptive of a person of known political and cultural attitudes and aspirations. When such credentials have been established, then it would be an added bonus if the applicant were also Christian!

It is important to remember, therefore, that the use of the terms 'Catholic' and 'Protestant' throughout this chapter is not a comment on confessional difference but rather a means of most easily identifying the major groups involved in, and affected by, the conflict.

Individuals and families in conflict situations can be socialized into violence in many ways. In Northern Ireland the influences of politics, demography, employment and education have all served to develop what might be termed a 'sectarian psyche'. The political factors have been comprehensively covered by others (Darby 1983: 15–31) and, since causes of particular conflicts are not a priority of this chapter, only brief reference will be made to them here.

Politics

When the Northern Ireland state was set up in 1921, it had a population of about 1.5 million people. Approximately 60 per cent were Protestant. They were also Unionist in the sense that they supported the strong links which the new political entity maintained with Great Britain. There remained a significant minority of 40 per cent who were Roman Catholic and Nationalist in the sense that they would have preferred to have been part of a united Ireland with Dublin as its capital. Although this representation is simplistic in the extreme, the seemingly mutually exclusive aspirations and identities depicted still form the basis of the apparently intractable conflict of today. In order to understand the continuing violence, it is essential to consider these original political priorities and decisions in terms of their current ramifications for families and individuals living within Northern Ireland. The concepts of separation, frustration and suspicion are seen as crucial factors in the development and maintenance of violence and a sectarian culture.

Demography

There has always been a fair degree of physical separation between Catholics and Protestants in Northern Ireland, especially in the larger cities. This has come about as a result of historical factors and of a perceived need for safety. It is also a consequence of government policy where wards, especially in Londonderry, were gerrymandered to maintain a Unionist majority in a predominantly Nationalist city. Nonetheless, in the years between 1921 and 1968, a significant increase in the degree of mixing between the two groups did occur. However, the onset of the most recent spate of violence in 1969 heralded what was described at the time as the biggest enforced movement of population in Europe since the Second World War. This again was precipitated by a combination of intimidation, suspicion and fear.

An analysis of the 1991 census reveals that about half the region's 1.5 million population lives in areas in which more than 90 per cent are Catholic or 90 per cent are Protestant. Fewer than 110,000 live in areas of roughly equal numbers of Catholics and Protestants. Of Belfast's 51 wards, 35 are at least 90 per cent one religion or the other. McKittrick (1993) has argued that this geographical separation means that people can, in a fashion that seems natural to them, lead lives of near apartheid and rarely, if ever, mix or socialize across the religious and demographic divide. Some 24 years of sectarian violence have served to deepen these existing divisions.

Obviously there is a perceived sense of safety in 'being among your own'. However, in terms of the use of violence, such separation of the major groups allows for the easy and clear identification of targets; the 1993 bombing of the Shankill Road, for example. It also means that any response can be similarly targeted at the other community; the shooting dead of seven people in the Greysteel bar at about the same time is a case in point. Both were sectarian acts carried out in the near certainty that only Catholics or Protestants would be affected on each occasion. Such blatant sectarian attacks have tended to be rare in Northern Ireland. It is more normal that some 'justification' is given for each killing. These explanations can take the form of the victim being 'accused' of being a member of the security forces or of one of the many paramilitary groups. It can simply be for doing manual work for the security forces, as was the case in 1992 with the killing of seven workers returning from such employment at Teebane in County Tyrone. As these 'tit-for-tat' murders continue, individuals and families are becoming increasingly fearful of working in, or travelling through, the wrong area. As a result, social segregation is becoming even more marked. It is quite likely that, when a community turns in on itself in this way, meeting less and less of the other community, they then have less knowledge of each other as individuals, families or human beings. This increases suspicion, fear and ultimately the potential for sectarian violence.

Concomitant with this rise in sectarian segregation has been an increase in a kind of siege mentality within the different communities. In many regions the sense of fear and threat has become almost palpable. This has permitted the paramilitaries to present themselves in the role of defending, and indeed policing, the 'ghettos'. This role seems to have had an attraction for the young people of the areas in terms of emulation and recruitment, this despite (or perhaps because of) the fact that some of the justice meted out has included killing, kneecapping, tarring, beatings and exile of members of the community for such crimes as minor theft, drug abuse or joyriding. It should not be surprising, therefore, that many of the young people of these areas become involved in the ambient violence, since many see it as exciting and legitimized. Quite apart from a commitment to any cause, such involvement can provide them with a status and esteem many of them would not otherwise experience.

Employment

People may be socialized into violence through frustration with their standard of living or quality of life. Both these concerns are likely to be critically affected by whether or not one has a job. In fact the Cameron Commission, reporting on the disturbances of 1968 and 1969, concluded that a sense of injustice felt by Catholics had been a major contributory factor to the violence. This sense of injustice was related to complaints about discrimination in employment and was one of the main reasons for the street protests of civil rights groups at that time.

Cormack et al. (1989) record that the official male unemployment rate in Northern Ireland in 1989 was 19.3 per cent. This global figure conceals several significant disparities. For example, the rates are much higher in the west of the country where Catholics form the majority. In addition, Catholic men are 2.6 times more likely to be unemployed than their Protestant counterparts. This difference has remained almost unchanged despite the creation of the most sophisticated Fair Employment Act in western Europe. Nowadays, this disparity is probably explained by the existence of what A. Gallagher (1991) has termed 'indirect discrimination'. These are employment practices that may have the effect of discriminating against Catholics but are not necessarily designed to so do. For example, research demonstrates that school leavers in Northern Ireland use informal networks (personal contacts for example) to gain employment. Because unemployment tends to be higher among Catholic families, Protestant school leavers tend to have more access to those in work and so are more successful in obtaining employment. The consequence is that, while unemployment rates are high in both communities, they tend to be higher in Catholic working-class areas. This has the effect of engendering a common sense of grievance throughout Northern Ireland, which is often perceived in Catholic communities in terms of discrimination and injustice. I believe that the importance of the unemployment issue has consistently been underestimated in terms of its contribution to violence. In some Catholic areas, for example, the unemployment rate is 30 per cent and whole families have never worked and may never be employed. In these areas, frustration and a sense of injustice are attributed to a political power base with which the majority have never identified. The situation is seen clearly as a sectarian issue. It makes little difference if this analysis is in fact correct. That it is perceived to be true exacerbates feelings of separation and suspicion and provides the paramilitaries with persuasive propaganda material.

Education

The family unit and education system are probably the strongest socializing forces in any society. One of the normal functions of a school system is to pass on to the next generation that which is culturally cherished by the present one. In Northern Ireland, schools are segregated along religious lines. Research (Darby et al. 1989: 15–18) suggests that in 71 per cent of schools all the children come from one of the two major groupings. The polarization of teachers is just as pronounced. There now exists a small number of integrated schools (about 17) but these cater for only a tiny percentage of the school population. My own research interest (Murray 1985) has been in the extent to which schools can influence the formation or reinforcement of national identity. To this end I have spent long periods observing and participating in both Catholic and Protestant schools. In all this time I have never observed one instance of the use of violence being encouraged. What was commonplace, however, was that each type of school tended to idealize and extol different aspects of culture and, indeed, history. As one teacher put it: 'In Catholic schools they give a biased version of history. Here, we teach it as it was.'

These different accounts of history may well be reinforced by the myths and legends learned in the home. A British army officer tells the story of apprehending a seven-year-old child for throwing stones at his patrol. When asked why he was doing it, the child replied: 'You bastards have been exploiting us for 300 years and we are not putting up with it any more!' If this account is true, it would seem that the child was simply parroting the logic of the adults in the community. The point is that children may engage in acts of violence they have come to realize are acceptable to, indeed reinforced by, their parents. The daily rituals and symbols carried out within the separate schools also highlight difference rather than commonality. This can sometimes be seen within the broad categories of Unionism or Nationalism. More often, however, it can be observed through the celebration of events such as Commonwealth Day or the Queen's Birthday, or flying the Union Jack in Protestant schools. In Catholic schools, celebrations of occasions such as first Communion, confirmation or Ash Wednesday are normal aspects of the school year. It is important to emphasize that such events, while common and natural, are also totally school-type specific. It would be inconceivable, for example, for the former list of events ever to be celebrated (or even recognized) in a Catholic school. In fact it can be argued that each demonstrates a cultural position that is anathema to the other.

I have argued elsewhere (Murray 1985: 77–90) that in Northern Irish schools culture is caught rather than taught. But, since the process takes place in separate and segregated establishments, emerging identities may be all the more strong and exclusive. Children, albeit unconsciously, are socialized into an awareness of difference and distinctiveness. In any democracy this should be both natural and desirable. In Northern Ireland, there tends to be a general inability to handle difference. Too often the display of symbols is seen in terms of provocation, and distinctiveness as being synonymous with divisiveness. This is exacerbated by the fact that when children from different schools meet each other, often the interaction takes place within a competitive context. On one occasion, for example, while I was conducting research in a Catholic primary school, the ten-year-olds' soccer team was playing the local Protestant school. One of the Catholic teachers present was getting quite animated and when I teased him about it, he replied darkly: 'There is more to this than just football.'

As already stated, schools are rarely accused of actually contributing to the violence, but they can be fairly criticized for not doing more to promote mutual understanding and tolerance. At the moment, it would seem that children in both types of school are being denied the means to understand each other's heritage. Nonetheless, within the formal curriculum of Northern Irish schools, programmes such as the Schools' Cultural Studies Programme, Education for Mutual Understanding (EMU) and Department of Education sponsored cross-community schemes are attempting to increase awareness among children. In fact, EMU is a compulsory cross-curricular component for all post-primary pupils. These programmes operate at the attitudinal level in trying to bring about increased tolerance. However, the problem with attitudes is that the more strongly we feel about something, the less likely we are to be tolerant of opposing views. We can all examine our consciences in this regard in terms of such issues as smoking, conservation, contraception, divorce, homosexuality or abortion. If we have problems with any of these, or indeed with any issue, then think how much more demanding it will be to expect tolerance from children whose fathers or mothers have been killed by Catholic or Protestant paramilitaries or the armed forces. Consider too how attitudes must have been affected by events such as Bloody Sunday, internment or the Enniskillen or Teebane bombings.

The attitudinal approach to increasing tolerance in schools and lessening violence afterwards is therefore a difficult and sometimes hazardous pursuit. On occasion the contact hypothesis has been employed to this end through such strategies as joint classes and cultural trails. These sometimes yield results but, especially when contact is superficial, they can do more harm than good. I can remember, for example, on the occasion of one such joint trail the Protestant children being less than amused when being confronted, in a Catholic area, with the graffiti: 'Boom, Boom goes Mountbatten'.

Similarly, Catholic children were less than impressed by a scrawled score line: 'Paras 13, Derry nil' — a farewell message from that particular regiment to the people of the city after the events of Bloody

Sunday, when 13 unarmed civilians had been killed by the army. I was convinced at that time that the children returned to their schools less tolerant than when they had left and that the potential for violence among and between them had increased rather than diminished. Despite the difficulties, it would seem that the Roman Catholic Church supports the attitudinal approach. This is perhaps inevitable since it accommodates both the desire for schools to play a part in reducing violence and the conviction that separate Catholic schools have a right to exist.

The Use of Violence

In normal class politics, politicians and policies tend to be influenced by an expanding or contracting middle ground. The relative performance of the Liberal Democrats in Britain in the recent past, for example, seems to have had a significant effect on the reactive measures adopted by the other two major parties. In Northern Ireland, however, it would seem that ethnic, rather than class, politics obtain. In this context of ethnic strife, politicians tend to be preoccupied mainly with the possibility of being outdone or outmanoeuvred by the more extreme individuals in their own or other parties. In addition, within the 'politics of violence' it may well be tantamount to political suicide to be perceived as 'getting soft' with the other side. There may well therefore be an inextricable link between ethnic violence and the development of political intransigence. Each act of violence causes hurt and fear. Fear engenders the demand for response and reassurance. Most often this demand is satisfied, at least in the short term, by the utterances of the more extreme politicians or the violent actions of the paramilitaries.

Conflict theorists have long been preoccupied with attempting to understand how it is that violence comes to be accepted as a means of settling conflict. However, it may well be that violence is inevitable in conflicts that involve minority and majority groupings. Generally the majority group, in terms of numbers, maintains power and social control. (In South Africa the arithmetic is different but the principle remains the same.) In Northern Ireland this control tended to be maintained by a degree of gerrymandering, voting restrictions and discrimination against Catholics in housing and employment. Such policies were not uncommon during the 50 years subsequent to the setting up of the state of Northern Ireland in 1921. Although the situation has now improved significantly, in 1969 such policies had become part of the social fabric to such a degree that when the civil rights movement took to the streets in non-violent protest, this in itself was seen by the institutional powers as an act of subversion against the state and was responded to accordingly. In phenomenological terms, the establishment perception of violence actually precipitated and reified that violence. It is possible that in any majority–minority conflict, the perceptions of the demonstration of legitimate grievance on the one hand and of unacceptable violence on the other may always be present. Actual violence may be the ultimate and inevitable development. Unfortunately, recent world events would seem to reinforce rather than contradict such a view.

At another level, it is extremely difficult for outside observers to understand, never mind condone, the use of violence. The temptation is therefore to attempt to impose objective, yet judgemental, analyses on the conflict. However, it is relatively easy to be tolerant and objective if one does not feel under threat and, at the moment, both communities in Northern Ireland feel themselves to be under threat to varying degrees. For the participants, therefore, violence may have become not only essential but also functional. It can, for example, be used at the outset as a form of defence, then attack and perhaps, ultimately, to achieve some political objective. It is essential to understand that most of the violence perpetrated in Northern Ireland is not mindless. This is an uncomfortable yet important observation and one with which those with a genuine concern for the region will have to come to terms. Too often in the past, the violence has been presented in terms of amoral 'godfathers' seducing young people into criminal acts. If this were the case, then solutions would be much easier to achieve. In terms of explanation or apportioning blame, such a depiction may well have a comforting ring to it. However, it is neither accurate nor helpful. There are many highly motivated people who are actively engaged in violence on both sides. Proof of this has been provided by events such as the hunger strikes, when activists were prepared to die for their principles, and, in fact, ten did. In the past, politicians, especially in England, have failed to appreciate the degree of conviction and the strength of commitment. Strategies for dealing with the conflict have been adversely affected as a result. This may explain the apparent difference in attitude between Britain and the Republic of Ireland towards the policy of internment. The latter maintains that while such a course of action may be effective in curtailing and containing 'godfather' criminals, it proves counterproductive in the treatment of individuals who may be highly committed to a cause. In fact such a policy may serve (and has served) to increase such commitment and motivation.

The question is often raised that since both Protestants and Catholics are Christians, a supposedly peace-loving faith or group of faiths, how does it come about that they are prepared to use violence in a religious conflict? As I have already said, I do not see the conflict in Northern Ireland as a predominantly religious one. But even if it were, this would not distinguish it from the many other altercations, which are extremely violent *precisely because* they arose as a result of religious difference and dissent. The point is that the extent to which people are prepared to resort to violence as a possible solution to conflict is largely determined by the strength of commitment and identity with their respective, and presumably conflicting, aspirations. For some people in Northern Ireland, an identity with either Britain or a united Ireland is the single most important facet of their lives. Indeed, for some, it is almost a *sine qua non*. It should not be surprising therefore that they will go to any lengths to protect and promote their respective identities, both for themselves and for their children. I believe that in the context of resorting to violence, no one would be any different. The only distinction between us would be the required strength of the threshold stimulus.

It is interesting that the violence emanating from each side over the years has tended to take different forms. Since the IRA is attempting to achieve a withdrawal of the British presence, then almost anything British, or that represents the establishment, is viewed as a legitimate target. Hence, government offices, banks, police and army barracks, and even buildings such as post offices have been attacked. In addition, their human targets more often than not wear uniforms. On the other hand, Loyalist paramilitaries have a dilemma in this regard. To attack such property or personnel would be an assault on the very establishment they purport to cherish and defend. They have therefore tended to concentrate their attentions on elements that can be identified with Nationalism and Catholicism. Therefore, targets have tended to be Catholic churches, Gaelic football grounds and the homes of Catholics and, indeed, Catholic civilians. For this reason, Loyalist violence often appears to be more sectarian than its IRA counterpart. However, it is becoming ever more debatable if this is in fact the case.

When one poses the question 'why is violence acceptable?', the answer is often 'because it achieves its desired effect'. In Northern Ireland, many Catholics argue that the Unionist government would not have granted them any concessions were it not for the original civil rights upheaval and the subsequent IRA violence. In fact, since the introduction of direct rule in 1972, the Catholic population has experienced significant improvements, especially in terms of housing and standard of living. Many Protestants, on the other hand, perceive this equalling-up process in terms of capitulation to the violence of Republican paramilitaries. They are right in the sense that the violence has certainly speeded up the process. This is especially important at the time of writing when it would appear that, for whatever reasons, the Unionist population is represented as being obstructive towards the highly publicized peace process. But it is simply articulating a genuine sense of outrage because, despite the protestations of the British government, Sinn Fein (the political wing of the Provisional Irish Republican Army) has almost succeeded in achieving a place at the conference table through its support for violent action. Examples of similar processes have been observed in places like Cyprus, Palestine, Kenya and many others. It is remarkable how quickly the use of violence has expedited the metamorphosis from 'terrorist' through 'freedom fighter' to 'statesman'.

Academics, however, have the responsibility of subjecting the concept of violence to intellectual scrutiny and perhaps advancing alternatives to it. Such a process will have little relevance for highly motivated and totally committed individuals on both sides who are actively engaged in a war. These people are likely to be convinced, not only of the justness of their own cause, but also of the immorality of the establishment opposing them. In addition, many of these participants may never have considered an alternative to violence as a means of achieving their goal, since they have no experience of any other method of negotiating. In this context, psychologists tell us that abused children are more likely than others to become abusers when they grow up. If this is so, then violence in Northern Ireland is likely to be self-perpetuating in the same way. For many people in the worst affected areas, violence is part of their everyday lives and can take the form of bomb scares or actual bombs, sectarian attacks, road blocks and checks. In addition, since the principal attempts to find a solution to the political problem to date have been military, then, *ipso facto*, violence becomes not only legitimized, but also an accepted part of the problem. Nonetheless, when people meet at conferences to consider topics such as 'violence as a means of resolving conflict', they tend to consider only the violence emanating from the paramilitaries engaged in the conflict and not that being perpetrated by the state forces confronting it. However, it has been the nature of armies to use violence and in

Northern Ireland this is manifest from time to time. Again it can take the form of house searches, verbal abuse, baiting and harassment at identity checks or, in its most extreme form, the events of Bloody Sunday. Even if such behaviour arises from laudable intentions, it is inevitable that it will precipitate reactive violence and, indeed, it often has. Thus, very rapidly the state's 'solutions' become part of the problem and violence may become cyclical and almost endemic.

Conclusion

This chapter attempts to raise several questions with regard to the use of violence as a means of resolving conflict. In the first place we must ask if it is invariably misguided and dysfunctional and, if not, then what causes or contexts can justify its use. It is often argued that violence is intrinsically undesirable. But in most cases such judgements are proffered by outside observers. They seem to arise from a conviction that their constructions of reality are more objective and detached and better than those of the people who are actively involved in the conflict. In academic terms this may well be the case, but in phenomenological terms we cannot comment on what others are doing unless we know what they think they are doing. It is only when we understand the motivation that we can hope to comprehend the use of violence.

O'Malley (1993) claims that violence is now perceived to be almost a constant in Northern Ireland. It is in fact the norm, only becoming abnormal when, as seemed to happen in late 1993, we lose the ability to control it. The priority now is how to manage or regulate the violence. However, the more effective we become at managing it, the less urgent becomes the task of determining and understanding its underlying causes. Consequently, there is less of an incentive to bargain or make concessions in order to achieve a lasting accommodation. I think the time is long overdue for priority to be given to attempts to understand violence instead of simply developing structures to control it.

One cannot undertake an analysis of violence in Northern Ireland without making it very clear that the vast majority of people there are not involved in it and, indeed, lead remarkably normal lives. It is just possible therefore that we might learn most about the concept of violence by studying conflict situations in which violence is absent!

6. A Very Modern War: Terror And Territory in Bosnia-Hercegovina[1]

Cornelia Sorabji

> It is a war over land, money and power; and its military aspects have more in common with warfare in the Middle Ages or with the Thirty Years' War than with World War Two.
>
> (Moore 1994b)

> The largest single element behind what has happened in Bosnia is the collapse of the Soviet Union and of the discipline that that exerted over the ancient hatreds in the old Yugoslavia. Once that discipline had disappeared, those ancient hatreds reappeared, and we began to see their consequences when the fighting occurred.
>
> (John Major, Hansard, 23 June 1993, col. 324)
> (Quoted in Malcolm 1994)

Both the methods and the motives of the current war in Bosnia-Hercegovina have frequently been depicted as ancient, medieval, throwbacks to a bygone era. In this chapter I concentrate on the early stages of the violence in 1992 when the largest territorial conquests were made, suggesting that while certain aspects of it are indeed reminiscent of ancient warfare, in essence the Bosnian case must be understood as a very modern war in which extreme levels of brutality, frequently perpetrated in face-to-face encounters and seemingly dependent on the whims and wishes of local perpetrators, are adapted to two peculiarly modern circumstances.[2]

First, the violence is central to an effort to alter local understandings of the abstract, and modern, category of 'nation'. Brutality is aimed at humiliating, terrorizing and killing the 'enemy' population in order to remove it from the territory, but also at transforming the assumptions held by both victims and perpetrators about the very nature of identity groups and boundaries in order to prevent any future return of the exiled population. I will argue that in this attempt the invocation of and assault on the concept of neighbourliness plays a central role, and that this concept is related to concepts about nations and peoples.

Secondly, violence occurs here within a world which, far from medieval, is characterized by assorted international agencies and organizations (UN, NATO, EC, UNHCR) and by sophisticated media communications systems. Given that an international public is watching and the international organs have been set up to deal with questions of global security, international relations and refugees, this world is not merely a distant backdrop against which warfare occurs, but an aspect of the war, utilized by military and political leaders in former Yugoslavia to further their own ends. In this twentieth-century context, 'ancient' methods have helped fuel the mistaken 'ancient motives' hypothesis, which holds that the conflict is the result of all but incomprehensible Balkan passions and will cease only when all parties are exhausted by it, an interpretation which militates against significant international military intervention and has been espoused by Radovan Karadzic, leader of the Bosnian Serbs, and easily adopted by those foreign voices and officials who have opposed such intervention.[3] (Croat leaders, with their sharp distinction betweem themselves as true Europeans and the Serbs as Eastern barbarians, have not adopted this rhetoric.) In considering the relationship between 'ancient' violence and international interpretations, I will focus in particular on the ways in which late-twentieth-century West European understandings of genocide as bureaucratic, standardized and total may be played to advantage by those pursuing ethnic purity and territorial conquest.

The conference from which the present collection springs posed itself the question of what makes violence acceptable and sustains war as an institution. In as much as ethnically pure territories could not have been created in Bosnia-Hercegovina without war and violence, at one level the answer to such a question is simply that violence may achieve results that cannot otherwise be achieved. What I hope this chapter may shed more useful light on is the way in which, once war and violence have already been launched, they may then sustain themselves.

The chapter focuses on Serb violence towards non-Serbs in Bosnia-

Hercegovina and the following caveats merit inclusion in more than the usual footnote. The chapter does not suggest that Serbs are the only perpetrators of war crimes or that non-Serbs are the only victims. It does not suggest that Serbs are naturally or culturally more bloodthirsty than their neighbours, or that all Serbs, or all young male Serbs, are actively engaged in the violence. It does not seek to apportion relative responsibility to 'the Serbs', 'the Croats', 'the Muslims' or 'the Bosnians'[4] *en masse*, or to suggest collective guilt on the part of any of these.

The Violence

Many studies have noted that the perpetrators and orchestrators of large-scale violence tend not to conceive of their acts as those of aggressors, but to imagine them in ways that diminish or do away with their own aggression. Thus Arendt points to how 'evil in the Third Reich had lost the quality by which most people recognize it — the quality of temptation' and Eichmann understood his central role in the extermination of Europe's Jews as a duty performed against his own personal inclinations and therefore justifiable (Arendt 1963: 150). Davis (1965) in her study of Catholic–Protestant collective violence in sixteenth-century France suggests atrocities were conceptualized as a way of restoring moral order, driving out the sinners and leaving the righteous free to continue their lives as, conceptually, they should have been and used to be. Srinivasan (1990) on 1984 Delhi points to the perpetrators' understanding that 'they are doing it to us', citing the Hindu woman who climbed into the luggage rack to escape marauding Sikhs, in spite of the fact that in front of her eyes it was Sikhs who were being dragged from the train and burnt by Hindus (see also Spencer 1992). These points should warn us that violence is rarely understood by those who practise it as unproblematically 'acceptable', but tends, on the contrary, to be surrounded by cultural elaborations and preparations, which focus the mind away from the idea of violence as a simple, natural expression of individual aggression. In former Yugoslavia, for example, the idea that they are doing it to us (or would if we let them) and that of restoring moral order — 'this land is historically and rightfully Serb/Croat, the Muslims are really Serbs/Croats' — are strongly evident and have been generously supported by the media. (The role of the domestic media in fostering the war is almost impossible to overestimate. See, for example, Lukovic 1993.) In focusing on extreme and personalized violence, I then wish to consider its functional role in the war rather than subscribe to any notion of a Balkan cultural affinity with, or conscious open embrace of, plain brutality. Indeed such a notion, implicit or explicit in much of the commentary on former Yugoslavia, has merely served to obscure analysis of events and thus to contribute indirectly to sustaining the war.

Methods, including the long-term siege and starvation of towns, the symbolic use of violence (such as carving crosses into living flesh), the targeting of religious buildings and personnel for death and destruction, pillaging, arson, rape, castration, forcing victims to scorn their own cultural traditions (like making Muslims urinate in the mosque) or pay homage to those of the perpetrator (for example, forcing non-Serbs to sing Serbian songs) and other similar acts have become all too familiar in the many accounts given by victims. Such accounts have horrified observers and the bewildering level of viciousness is one reason for the representation of the violence as ancient. A second lies in the fact that, as well as impersonal killing and destruction accomplished by unknown military representatives of the opposing force, this war has seen a high level of violence perpetrated between neighbours, colleagues and erstwhile friends — personal acquaintances rather than impersonal representatives. A third and important factor lies in the local variations of scale and style of violence. The much-publicized accounts of rape, for example, include cases of women being assaulted in their own homes, in military headquarters, in camps, on the road to exile, allowed to buy their way free with Deutschmarks, released after impregnation and the possibility of terminating the pregnancy had passed, killed after being assaulted, assaulted once, assaulted continuously, told their perpetrators had been ordered to rape, told they had to bear Serbian children, raped as part of a supposed interrogation, told that their men were raping Serbian women, or told nothing. Castration of inmates has been reported in at least one camp, but not in others; different towns and villages have been physically destroyed to differing degrees. The variety of suffering on the part of victims from different parts of Bosnia-Hercegovina, and of statements on the part of perpetrators, contributes to an interpretation in which this sort of violence is the disorganized behaviour of random extremists who are out of control. This appearance of disorder has, I suggest, important implications for popular perspectives taken in West European countries.

The View from Afar and the Question of Genocide

Moore (1994a) aptly notes that 'in contrast to the classic relationships between the major powers and their Balkan allies, in which the former used the latter as pawns in a broader game, in the current conflict the Balkan armies held the initiative, sending international diplomats scurrying about in futile efforts at damage control.' It may further be noted that international scurrying, while perhaps futile from the point of view of those doing it, has been relied and calculated upon by former Yugoslav leaders as a support to their aims. As Vasic (1993) forcefully notes:

> The Geneva talks have become an industry for the legitimization of war crime profits. ... The final sad conclusion is that there never was any peace process. It was a sham from the very beginning ... a diversionary side-show mounted by two card sharps [i.e. Presidents Milosevic and Tudjman of Serbia and Croatia] talking peace in Geneva and land in Bosnia. It was a simple blackmail: make us happy or we'll kill another 10,000 people.

'Ethnic cleansing' is 'interdisciplinary' (Loza 1993) and employs military, political, diplomatic and psychological methods and a range of tools including the media, international diplomacy, aid agencies and UN troops to remove the unwanted people from the targeted area. In this sense ethnic cleansing could be understood to include not just actual events of physical violence but also, for example, the endless drawing up of maps in Geneva, the imposition of a UN arms embargo on Bosnia-Hercegovina, the succession of unfulfilled threats issued by the international community to the Bosnian Serb leadership, the EC's early legitimation of nationalist leaders and principles — which, while obviously damaging to Muslims, also pushed previously uncommitted Bosnian Serbs and Croats towards opting for membership of ethnically exclusive statelets (see Sorabji 1993) — and the UNHCR's difficult decision to evacuate endangered civilians from hostile areas. From this perspective it is hard not to conclude that the organs of the international community have, in culpable and non-culpable ways, implicated themselves in the project of cleansing.

Less clear is the relationship of the newly-coined term 'ethnic cleansing' with the older one 'genocide', devised in the context of the Second World War. Writing of events in former Yugoslavia, the UN's own Commission of Experts notes that: 'Those practices constitute crimes against humanity and can be assimilated to specific war crimes.' Furthermore, such acts could also fall within the meaning of the

Genocide Convention (January 1993 Interim Report of the Commission of Experts Established Pursuant to Security Council Resolution 780, 1992). Yet overall there is a reluctance to perceive similarity and it is suggested here that the 'ancient' texture of the violence is one factor that may obscure the question.

For most of us alive today the Second World War is *the* war and the Final Solution is *the* genocide and the thing that must never be allowed to happen again. It is through this lens, among others, that Bosnia-Hercegovina has been viewed and at least some of the early debate over military intervention revolved around the question of how similar or different Serbian ethnic cleansing was to the Nazi Holocaust. Overall, the tendency was to highlight the (unquestionable) differences rather than the painful similarities. Thus, for example, our predominant image of the Holocaust is drawn from the later stages in which all Jews were to be systematically exterminated and the prior, if extremely slim, possibility of emigration to Palestine or elsewhere was removed. In Bosnia-Hercegovina more people have been displaced than murdered, yet it should be noted that, according to the Genocide Convention, genocide means 'acts committed with intent to destroy, in whole or in part, a national, ethnical, racial or religious group'. Totality is not a necessary criterion of genocide, rather it is the popular view of genocide, derived from the Second World War, which emphasizes totality, thus diminishing the Balkan case as less-than-genocide but also, simultaneously, diminishing the specific and particular horror of the Nazi genocide.

The central place the Second World War holds in our understanding of war became especially apparent over the discovery of Omarska, Manjaca, Trnopolje and other camps in Serbian-held Bosnia. Reports that these were 'concentration camps' were in many quarters rejected in spite of the fact that they were precisely places in which unarmed civilians, mainly young and middle-aged males, but also elderly people and women, were concentrated. People of the 'wrong' identity were rounded up, transported there, held against their will, forced to labour, starved, beaten, tortured and, many of them, killed — as were some of the 'right' ethnic/national identity but the wrong ideological persuasion. Omarska, for example, contained some Serb prisoners who refused to cooperate with the new Serbian statelet. Objections to the term 'concentration camp' appear to have been based around the fact that these were not like Nazi concentration camps. For example, 'Some [US] officials have suggested that since the existence of systematic death camps cannot be proven, the camps should not be referred to as "concentration camps" and that, therefore, the conditions in the camps are not as poor as press reports would lead the public to believe' (*Helsinki Watch* 1992: 131).

Not everyone in these camps was exterminated: there were no gas chambers and no tattooing, and records of inmates were kept less fastidiously than in Nazi camps. These camps were not those of the Holocaust — they were less standardized and orderly than the gas chambers disguised as showers, the gold teeth and spectacles confiscated and saved, the huge camps and railway lines to them, the 'medical' experiments and the bureaucratization of genocide. The Bosnian case then appears relatively unbureacratic and decentralized, yet there remains the question of whether destandardization and disorderliness imply a lack of organization, or merely organization of a different type in a different political, historical and cultural setting. It is possible to see the Bosnian case, in a loose sense, as a sort of franchise organization in which local mayors and military commanders are free to organize and encourage violence in their own regions in their own particular ways, and into which some 'freelances' inject their own individual sadistic methods and imaginations. Rather than an organization of specific violent techniques, this suggests the organization of a context in which people are enabled to inflict whatever disorganized tortures they may dream up.[5] The context, however, *is* organized, as shown for example by the Celinac curfew order, or the mayor of Prijedor on the establishment of the camps (Harriman 1993), or the division of Omarska prisoners into three categories on the basis of pre-written lists. It is this scenario that makes sense of the answer given by a Muslim ex-inmate of Omarska when asked whether he thought the torture was organized — an astonished gaze, shake of the head and the reply: 'Anyone could come there and do whatever they liked.' Or of another ex-inmate's comment that 'Omarska camp was open for all those Serbian volunteers who had someone of "their own" in it, some captive on whom they wanted to vent their rage. These were usually first [next-door] neighbours (*prve komsije*) of the people inside' (Hukanovic 1993: 56).

From this perspective, the disorganized look of the violence, the regional variation, the assortment of acts and the assortment of explanations perpetrators give to victims about their acts need not imply the completely uncoordinated and separate acts of 'extremists' or archaic 'warlords', but rather the predictable outcome of a franchise organization in which the general aims have been set from the top and the specific details left to local initiative. However, the central place of the Final Solution in European minds tends to make a messy, unstandardized, unbureaucratic approach to mass murder appear less genocidal and, in this respect, the 'ancient' quality of the violence serves to alleviate some of the constraint that might otherwise pressure the modern world to 'undertake to prevent and punish this crime' (Genocide Convention).

Nations, *Narods* and Neighbours

If the popular perception of what genocide is has proved to be of functional use to ethnic cleansers, it may also be said that the widely held view of what nations and nationalism are — nations as natural cultural and political entities and a sense that it is not surprising that such groups want a state or territory to themselves — have provided backup. Numerous writers (notably Gellner 1983; Anderson 1983; Hobsbawm and Ranger 1992) have examined the emergence of such a view in the modern world, stressing that nations are far from natural or age-old, but equally stressing the firm grip such an idea has on the modern imagination. It is partly this grip, I suggest, that has made it easy for West Europeans, themselves long ago divided into nation states, to understand Bosnia-Hercegovina so wholly within this idiom, tending to ignore the possible existence and salience of other, non-national, collectivities. Of the variations on this theme evident in international commentary, the most pronounced has been that which presents pre-war Bosnia-Hercegovina simply as a region inhabited by three separate nations — Muslim, Serb and Croat. Occasionally the religious sound of the title 'Muslims' has led to confusion as to whether the Muslims are 'really', i.e. nationally, Serbs or Croats. In either case the existence of distinct nations is taken to be the most important model for understanding pre-war society and in both there lies some implicit support for the 'world view' of Serb and Croat national expansionists. On the other hand, observers with a sympathy for Bosnian unity have tended to use the other side of the same coin in their arguments. In this variation one string of the argument for a united Bosnian state is that Muslims, Serbs and Croats are neither ethnically nor nationally distinct.

All these versions miss the subtle complexity of the pre-war Bosnian understanding of collective identity by assimilating it into Western European notions of nationhood and ethnicity. Part of the problem lies in the very use of the English word 'nation' as a translation of the Serbo-Croat term '*narod*'. In ordinary everyday speech '*narod*' means 'people' or 'a people'. Thus one can say 'people were angry', 'the people enjoyed themselves' or 'the Japanese are an intelligent people'. In comparison with the word '*svijet*' ('world' or 'people') and more so with the word '*ljudi*' ('men' or 'people'), '*narod*' more strongly implies some sort of collectivity. This is why it fits well with situations of anger or enjoyment, where those experiencing the emotion are conceived of as a group rather than as many angry or happy individuals. At the same time, however, in official parlance the word '*narod*' formed part of the Yugoslav state's doctrine on national identity. Six official '*narodi*' were recognized — Serb, Croat, Slovene, Macedonian, Montenegrin and Muslim (the latter achieved over a period of time culminating in the 1974 constitution's recognition of the Muslims as a *narod* equal to the other five). These six should not be seen as pure inventions of the socialist state (in spite of some Serb and Croat propaganda to the contrary in respect of the Muslims). It was not necessary, for example, to send ethnologists to discover/create nations out of the human fodder of the region in the way Lenin sent ethnologists to Soviet Central Asia in the 1920s. Divisions between these groups were recognized prior to socialist rule and under it the enshrinement of the *narodi* in the Constitution increased their salience to former Yugoslavs.

Both the everyday speech view and the official view then formed part of the mental repertoire of ordinary citizens of Bosnia-Hercegovina. The term 'official view' is not then intended to imply that it was rejected or unrecognized by the public. However, the official view of the *narodi* did not exactly correspond to the everyday speech view and two aspects of the difference are important here. First, the everyday view of *narod* as 'a people' would include groups such as the Albanians, Hungarians, Italians and others. These groups were not officially classed as *narodi* but as lesser '*narodnosti*', the professed reason being that they did not live wholly or mainly within the borders of Yugoslavia — most Albanians live in Albania, most Hungarians in Hungary and so forth. Thus the official view posed *narod* as something specific to the Yugoslav condition, dependent not only on factors such as language, religion or history, but also on geographical position in relation to the borders of the state.[6]

Secondly, the official view cast all Yugoslavia's Serbs as one *narod*, all its Croats as one *narod* and all its Muslims as one *narod*. In practice, however, Serbian Serbs and Croatian Croats tended to view Bosnian Serbs and Bosnian Croats as different from themselves, as culturally more backward and primitive. In the Muslim case the order was reversed with Bosnian Muslims seeing themselves as superior to their allegedly primitive Muslim co-*narod* members in the Sandzak region of Serbia and Montenegro. To those outside Bosnia the republic's inhabitants were '*Bosanci*' — Bosnians — be they Muslim, Serb or Croat. Within Bosnia people understood the division between Muslim, Serb and Croat *narods*, but nevertheless distinguished Bosnian members of these *narods* from non-Bosnian members. Thus Bosnian Croats were frequently referred to as '*Katolici*' (Catholics), as opposed to '*Croats*', a term largely reserved for Croatian Croats. In relation to the other republics they understood themselves as *Bosanci* and, in defiance of the widespread view of Bosnia as primitive, claimed allegedly Bosnian virtues such as spontaneity, generosity, warmth, easy-goingness, good cuisine and beautiful women. In as much as *narod* meant a collectivity of people or 'a people', Bosnia and Bosnians were then so seen both by outsiders and by themselves.

Clearly, this view was not explicitly catered for by the official view of *narod*, yet it is important to note that the whole idea of *narod* in official parlance already, if ambiguously and with fluctuating emphasis, included the idea of 'Brotherhood and Unity', the Yugoslav motto. In the early days of socialist Yugoslavia there was a hope that a Yugoslav nation was in the making. At the eighth League of Communists Congress in 1964, Tito disowned this idea and promoted one of organic Yugoslavism or Yugoslav Socialist Patriotism, in which the various different *narods* lived together in harmony (see Ramet 1992: 51–4). The official view of *narod* therefore already included that of the brotherhood and unity of the Yugoslav *narodi*, *narodnosti* and ethnic groups (*etnicke grupe*). In this sense it was a specifically Yugoslav concept not readily reducible to an exclusivist West European notion of nation.

What is being suggested is that the notion of *narod* is more multifaceted than that of nation and that Bosnian citizens understood it in different ways in different contexts. At least three understandings are evident:

- Muslim, Serb and Croat are separate *narods*.
- In as much as *narod* means people or a people, all Bosnians jointly constitute such a people/*narod*.
- Even the separate identities of Muslim, Serb and Croat *narods* are implicitly joined through and predicated on organic Yugoslavism, brotherhood and unity.

The creation and, more importantly, the maintenance of ethnically pure territories thus requires that the first view of *narod* be retained and reinforced while all others are jettisoned. In this project personalized violence involving the killing of neighbour by neighbour, colleague by colleague and friend by friend acquires a particular importance. The idea of 'neighbourliness' formed a dominant idiom within which people understood and spoke of life and social relations. Neighbours are deemed to be important socially, ritually and economically. They attend your rituals, lend you money, housewives have coffee together most days, if not twice a day. In Sarajevo under socialism the idea of the neighbourhood as a safe area away from the town centre and in which one could be freer in one's speech gave it a special relevance. Indeed 'neighbour' is an idiom for good relations in general, and those who live somewhat outside one's geographical domain can nevertheless be defined as neighbours if one exchanges with and is on good terms with them. The expression 'first neighbour' (*prva komsija*) implies not only geographical closeness, but deeper relations of trust, affection and exchange. In some respects, then, the concept of 'neighbours' takes the place of that of 'brothers' in other societies, standing for the strongest possible moral and social tie outside the immediate family. This is particularly (though not exclusively) true of Muslims, who have noticeably less interest in and knowledge about their genealogies and blood kin than the Croats and, especially, the Serbs. In social life, then, neighbours were of extreme importance. Conceptually the relation of neighbourliness was one which, *par excellence*, represented warmth, trust and interconnectedness. At the same time, however, as people with whom one exchanges goods, words and ritual services, neighbours are conceived of as different, as 'others' with whom relations of exchange are appropriate. In this respect the concept of 'neighbours' mirrored that of *narodi*: the *narod*s were separate, but even their separate existence was predicated upon their interconnectedness; neighbours were warm, trusting and united, but at another level this unity was predicated upon difference.

In this context the May 1991 attempt by Nikola Koljevic, deputy leader of the Bosnian Serbs, to reassure Muslims in Bileca that 'a neighbour is more than a brother' (i.e. 'you Muslims are more to us than our own, Serb, brothers') may be seen as a statement invoking the commonplace understanding of the extreme value placed on neighbourliness, trust and interconnectedness and, simultaneously, as a statement stressing the difference and distance between Muslims and Serbs.

In the war itself the use of neighbour-to-neighbour violence privileges the latter connotation of 'neighbours', that of difference and distance, and is, I suggest, intimately connected with the attempt to give priority to the understanding of '*narod*' as an exclusive category over other understandings of the term.

Terror and Memory

Many of the camps have been makeshift, established in local schools, factories, mining complexes and so on. Many of the torturers and murderers were neighbours, colleagues or friends. Clearly this is in part a question of practicality. Camps must be established quickly and the closest building to hand may be the easiest. Local people are on hand to do the torturing and murder and it would be hard to draft in enough outside forces to do the job alone.[7] However, the personal nature of the violence and the fact that it is frequently accomplished in physical spaces and settings all too well known to the victims is in itself an essential factor in the success of ethnic cleansing. Neighbours and neighbourhoods, spaces and places, are deeply imbued with memories of shared lives and of notions of what Bosnia is. By transforming neighbours from economically, ritually and socially essential relations into torturers and killers, by transforming neighbourhoods into slaughtering fields, by transforming the very school one used to attend into a camp where an ex-classmate or pupil now holds you captive, people's images of the lives they used to lead, the meaning of places and the meaning of 'neighbours', which is intimately associated with the meaning of Bosnia and of some connotations of *narod*, may be distorted or destroyed.

This erasure of meanings and memories is discernible in the written and verbal accounts of those who have been held captive, or have been tortured. Although diverse in other respects, most accounts tend to be notable for the flatness and simplicity of the narrative. One is told what happened and in what order in straightforward terms. Secondly, they are notable for their attention to details of time and place. One is invariably told the day and date of the events, and frequently given the exact timings of successive events — 'at 4.30 in the morning three uniformed men arrived at our house' … 'at 11.00 in the morning we went to X's house' … 'that afternoon at 2.30 A.B. came over and took my brother away'. The events that occurred in X's house, on the bridge, at the headquarters, outside the mosque are all noted with special attention to place. Before terror sweeps down people will have heard rumours from other villages or towns and will often have been involved in 'negotiations' over the surrendering of what weapons they have (for the most part only pistols and hunting rifles). When violence hits it hits right in the home and the neighbourhood. In reading or listening to accounts it is hard to avoid the conclusion that what people are telling is a simple, hardened but dominant last image of the place they used to know, an image of a completely other place, transformed from the one they previously knew.

Personalized terror and violence serve the purposes of frightening people into leaving and frightening them into never wanting to return, even if their homes are still standing for them to return to. But what they may also do is erase old memories and images, creating a 'blank space' in a way that leads some (though not all) victims to reinterpret the past. For some of the cleansed the process of cleansing leads not to the conclusion that 'we all used to live happily together but then they changed', but to the conclusion that 'they always hated us; they were always waiting; Bosnia and neighbourliness was an illusion'. Old places and old neighbours are not what they used to be and old ideas and images are erased and retrospectively rebuilt and replaced.

On the other side of the violent line, the perpetrators' side, there is also erasure and the creation of blank space, for *both* sides, not only the victims, must be made to forget how things used to be and be understood. In the case of the self-proclaimed *Republika Srpska* (Serbian-held Bosnia), images of violence, feelings of fear and/or guilt, and the fact that there seems to be no going back are likely to contribute to people's capacity to forget, but they are also given practical and symbolic help. For example, by decree of 'parliament' place names are changed to eradicate the word 'Bosnia' and hints of any Turkish past. Bosanski Petrovac becomes Petrovac, Bosanski Novi becomes Novi Grad, Kulen Vakuf becomes Serbobran, the Banja Luka street named Skender Kulenovic after the famous Muslim poet becomes Stojanka Majka Knezopolke after the Serbian woman honoured in Kulenovic's verse, the East Bosnia Army Corps becomes the Drina Corps, even the famous dish *Bosanski Lonac* (Bosnian Hotpot) is reputedly in certain areas now referred to as *Srpski Lonac*. Such examples are mirrored in Croat-held territory. Thus Gornji Vakuf becomes Uskoplje and the University of Mostar, so named after the famous bridge destroyed by Croat forces, becomes the Croatian University of Mostar. In Bosnian-held Sarajevo the street named Stojanka Majka Knezopolke recently became Skender Kulenovic in a reversal of the Banja Luka story. In this effort to erase the old and erect the new, even time itself may be redefined. Thus Mayor Vucurevic of Trebinje, when asked how he could destroy the beautiful old town of Dubrovnik, replied that it did not matter because 'we will build a new, more beautiful and older one'.

Conclusion

Such attempts, conscious and unconscious, to erase the past, its images, memories and meanings and to build new ones in their place seem likely to sustain and produce more violence. In Petrovac where the presence of Bosnians reminds inhabitants that this was once Bosanski Petrovac it must be asked for how long the remaining Bosnians will be tolerated. After the first and most publicized wave of cleansing from the north-western town of Banja Luka, many more non-Serbs were later driven out or killed. The fact that a region or town is safely held militarily and politically does not make the remnants of its ethnically 'impure' population any the more acceptable to the ethnically 'pure' majority, since these remnants are a reminder of old ideas of neighbourliness, *narod*s and Bosnia.

It may further be noted that while a franchise and freelance approach to mass terror and murder may have proved a highly effective method of ethnic cleansing, it provides little guarantee of producing a peaceful and orderly new state in the future. The encouragement of individual violence and the proliferation of local leaders suggest that the self-proclaimed *Republika Srpska*, if fully ethnically 'purified', may later become riven with interregional Serb–Serb conflicts (as hinted by the disturbances in Banja Luka in late 1993), which could only be suppressed through authoritarian 'state' violence and terror.[8]

'Ethnic cleansing' clearly aims at removing people through methods that include murder, terror, torture and destruction. Beyond this I have tried to suggest that the seemingly ancient texture of the violence, and in particular its vicious, personal and apparently spontaneous quality, is directed at altering the beliefs and mental categories of former Yugoslavs in an effort to ensure that survivors of the initial cleansing will not later return.

The impact of terror and violence applied in familiar surroundings by familiar people (actual neighbours or conceptual 'neighbours') and in personalized ways helps to obliterate prior assumptions and memories about *narod*s, neighbours and Bosnia-Hercegovina, and to obliterate them for both victims and perpetrators. Although ancient imagery and symbolism are frequently invoked, their aim may in part be seen as the fostering of an arguably more 'modern' notion of nation as exclusive and overridingly important.

I have also tried to emphasize that the war occurs within a particular, modern international setting, which is itself an aspect of the conflict. Former Yugoslavia is not an oasis of ancient madness disconnected from the modern world: on the contrary, its political and military leaders calculate upon the actions of the international community in attempting to further their war aims. In as much as seemingly ancient violence ill fits the popular Second World War-derived notion of genocide, destandardized, 'disorderly' organization and personalized brutality, while aimed primarily at affecting the target population, also serve to delude and to foster the self-delusions of the outside world.

Notes

1. Many people have made helpful comments on aspects of this chapter but I would like to record a particular debt to Milos Vasic and Tihomir Loza, both of whom have provided me with examples and from both of whom I have learned much in the course of general conversations. Naturally, neither is responsible for the opinions expressed here, which are mine alone.
2. This chapter focuses on the face-to-face atrocities which have reinforced a view of the war as motivated by ancient hatreds. It should therefore be stressed that this type of personalized violence is but one in an array including heavy shelling from a distance, use of anti-aircraft missiles against ground targets, and of snipers.
3. A frequently expressed notion is that when the communist lid was eventually removed, the boiling pot of nationalist hatreds erupted. This view of communism and nationalism as simple forces, the one removed to reveal the other fully formed and ready to explode, ignores the ways in which the experience of communism and the process of its removal themselves contributed to nationalist feeling. Part of this contribution was one of fear which, rather than hatred, provided the initial psychological grounding for the war (Sorabji 1994: 125–6).
4. The use of the term Bosnians requires some explanation. In pre-war Bosnia-Hercegovina, the term Bosnians referred to the inhabitants of the republic, Muslim, Serb, Croat, Jewish, Romi and so forth. Since the war the term 'Bosnian' has come to denote those inhabitants, mainly Muslim but also Serb, Croat and Jewish, who support the idea of a non-ethnically-based state. What new referent the term may acquire in the future remains to be seen.
5. The reasons why rape is such a common act of violence cannot be discussed here but it is interesting to note that the idea of sexual assaults on women as a nation–nation crime, as well as a male–female one, had previously been fostered in the context of mid-1980s Kosovo (see Meznaric and Zlatkovic 1991). This point has an obvious bearing on the question of the organization of a context in which violent acts may be committed.

6. The ideological difficulties involved in reconciling the national question with the advance of the universal socialist man were apparent throughout the multi-ethnic communist bloc and reflected in the nationalities' policies of the various states.
7. Although outside forces from Serbia have been used in face-to-face ethnic cleansing, particularly in the early assaults on Eastern Bosnia.
8. If internal conflict is the probable eventual outcome within a self-proclaimed *Republika Srpska* it is also true to say that the experience of diverse Bosnians (i.e. mainly Muslims but also others within the multi-ethnic camp) are diverse and may produce diverse aspirations in future. In Bosnian-held Sarajevo, inhabitants have been shelled, starved and snipered but not incarcerated or cleansed face-to-face. In Mostar, Muslims have been shelled, starved and shot at by Croats. In addition to the diversity of war experience it is also true that pre-war Bosnia-Hercegovina, like the whole of Yugoslavia, was characterized by regional diversity of customs. It is therefore impossible to illuminate an entire picture of how Bosnians do or will understand their experience and identity, or their identity in future.

Part II

ENHANCING MOTIVATION

Introduction

The chapters in this section explore some of the pervasive cultural factors that act to make war acceptable to those involved or likely to be involved. This includes not only the combatants, but the support services and indeed other sectors of the population as a whole. Because the institution of war involves role-differentiation — the behaviour not only of soldiers, but also of doctors and nurses, munitions and transport workers, civil servants, policemen and many others is determined in large part by the duties accompanying the roles each occupies in the institution of war — they must also accept all that war brings with it. Furthermore, war usually involves conflict between complex societies, each of which consists of overlapping groups with some similar and some conflicting interests. The acceptability of war thus depends on the society being seen as a beneficial whole.

Thus there are two main issues with which we are concerned here — factors that enhance the perception of the state or group as unitary and beneficent, and those that promote the willingness of individuals to subordinate their own interests to those of the group to which they belong. These goals are, of course, closely interrelated: indeed in an earlier chapter, M. Lewis (Chapter 1) described how the legalistic philosophers of Warring States China saw them as sequential. Rewards for heroism and punishment for cowardice could lead to victory in battle and therefore to prosperity. The people would thus see private benefit in the preservation of the prosperous conditions created, and would thus come to love their rulers and do their duty so that order should be preserved.

In more recent times societies differ markedly in norms and values associated with both interpersonal violence and war. Amongst industrial societies, some have a long history of belligerence; some, such as Switzerland, of neutrality. Some, such as Sweden, have changed from being aggressive to peace-loving. Three relatively peaceful pre-state societies described by T. Gibson, T. Gregor and C. Robarchek (in Haas 1990) all possess an anti-violent value system. Quarrelling, anger and violence are stigmatized; generosity, gentleness and conflict-avoidance accorded prestige. These three societies also tend to be egalitarian, attempts to dominate others being frowned upon: this is associated with dependence on the group rather than on the individuals in it and thus, by Western standards, the impoverishment of personal relationships (see also Goody 1991). In such societies at least, then, there seems to be a relation between early socialization and the incidence of war. Whilst one must expect such relations to be much diluted in more complex societies, in every society individuals differ in their evaluation of war and in their responsiveness to war propaganda: socialization and educational practices are likely to play a role in this. Indeed the distinction between education and propaganda for war can be a fine one. In Chapter 1, M. Lewis described the Confucian prescription for education as including martial elements. A number of other chapters touched on the way in which socialization and education prepare individuals for warfare (for example, G. Lewis, Drucker-Brown and Papadakis in Chapters 2, 3 and 4 respectively). In this section, Smith and Siegel (Chapter 7) discuss the role of socialization and education in promoting or diminishing interpersonal aggressive tendencies and susceptibility to war propaganda in North America (see also Feshbach, Chapter 11).

An everyday issue that promotes the acceptability of war lies in the 'warisms' we use in everyday speech. Fussell (1975) has described the 'high diction' that, at least until 1914–18, concealed the true reality of war. Thus the 'dead' were referred to as the 'fallen', 'not to complain' was to be 'manly' and so on. Newspapers still abound in war metaphors — 'going over the top', 'getting dug in', 'outflanking' and so on. New 'high tech' military euphemisms, such as 'collateral damage' and 'smart bombs' seemed to proliferate during the Gulf War, perhaps as a corollary to the 'Star Wars-like' and impersonal nature of modern warfare. It is hardly surprising, given the time that they were written, that the Psalms use many war metaphors, but Christian hymns are almost their equal in that respect. In due course, warisms could be as demoded as sexisms.

Many of those who have gone to war have seen the coming conflict as releasing them from the dependency and infirmity of the older generation, and as an occasion where their virility can be fulfilled in *Kameradschaft* (for example M. Lewis, Chapter 1; Murray, Chapter 5; Mosse, Chapter 9). However, although the difference between the expectations of the new recruit and the reality of war is often devastating (Brodie 1990), once war is over, memories of its horror are often less durable than those of comradeship and release from the tameness of civilian life. Mosse (1990) has described how the memory of the world wars has been reshaped by literature, and Winter (1991) has made a similar point about the postcards issued during the First World War and films made subsequently. In this volume Mosse takes up the particular issue of the glorification of aviation as epitomized in the phrase 'Knights of the Air', and its effects even on the infantry in the First World War. He also traces relations between combat flying, flying as a sport, and mountaineering, which are in some ways reminiscent of the relations between killing in war and hunting in New Guinea as described by G. Lewis (Chapter 2).

How potent such issues may be no doubt depends on other aspects of the culture. M. Lewis (Chapter 1) records how militaristic images persisted in Chinese literature long after the Warring States period, into a period when the only wars were on distant frontiers with hostile nomads. Nevertheless, the military continued to serve as a model for secret societies, which formed the basis of rebellions. Present-day 'terrorist' groups seek legitimacy for their actions by borrowing militaristic models and terminology (Watson, Chapter 12). The use of flags, parades and other devices for maintaining long-term group solidarity was emphasized by Papadakis (Chapter 4). Such issues may also encourage recruiting in times of war: in the early days of the 1939–45 war, UK government advertisements asking for volunteers for aircrew training utilized just the themes stressed by Mosse, and were certainly effective.

But something more potent may be needed when it comes to actual combat. This is especially the case when killers can see victims — either through the sights of a sniper's rifle or at close range — and when fear is likely to become paramount. In some cases, a fighting mood may be aroused or augmented by special devices. To take an example from pre-state warfare, young Yanamamo warriors 'put on rage' as a tactic, and use drugs to induce a fighting mood (Chagnon 1990), and in some New Guinea groups fighting takes place only when warriors have achieved a trance-like state (G. Lewis, Chapter 2). And soldiers in the 1914–18 war were given hate-circulars to arouse their antipathy to the enemy, though apparently with little effect (Tillyard 1962). Many of the devices used in this context have been mentioned in earlier chapters. For instance G. Lewis (Chapter 2) described the rituals used to prepare men for battle in New Guinea, and M. Lewis (Chapter 1), writing of early China, stressed the importance of uniforms as indicators of status and as endowing a military mystique, discipline, music in battle, the creation of a military macho image, faith in the commander and so on. The break-up of Yugoslavia and the Troubles in Northern Ireland have brought to public notice the incredible callousness and cruelty that are possible in war. Sorabji (Chapter 6) and Tromp (Chapter 8) discuss how such behaviour can be seen to be normative under certain conditions in war. Tromp emphasizes the creation of a culture of war with alienation from normal values, normal social controls and military discipline.

War involves organization, and combatants must undertake actions unlikely to bring them personal benefit. This requires leadership, and the role of the leader has been emphasized in a number of the chapters in this volume. M. Lewis (Chapter 1) recorded that, in some Warring States' treatises, faith in the commander was seen as the key to unified action, and discipline resulting from extensive training as essential.

But, as we have seen, propaganda must be aimed not only at the soldiers in the field but at the population as a whole, and must aim to produce group solidarity, a sense of common purpose and denigration of the enemy. It is a commonplace amongst social psychologists that individuals who see themselves as members of a group almost inevitably see the members of their own group as superior to those of other groups. How far they see themselves as members of a group depends on a number of factors, including the extent to which they see the group members as interdependent with respect to common goals or the means to attain those goals, and as subject to a common fate, and see the group as a distinctive social unit (Rabbie 1991). It is precisely on these issues that much propaganda depends. It emphasizes common benefits in victory, a common fate with defeat, and the need for cooperative effort. It also reinforces group unity with symbols (like flags and national anthems), parades and speeches.

While some social psychologists (for example Tajfel and Turner 1979) see relations between groups as essentially competitive, others argue that this depends on the psychological orientations in the group or nation and on the nature of the interdependence structure between them (Rabbie 1989). War propaganda accentuates the competitive element by representing the enemy as dangerous, evil and even subhuman, and by inducing an element of fear. Propaganda and misinformation can operate at the grassroots of society to create new cultural and symbolic boundaries in a hitherto unified community. In a context of emergent conflict, rumour of atrocities and anticipation of imminent attack further intensify the process of division and opposition, by the 'demonization' of the enemy (Watson, Chapter 12). The processes involved in the propagation of an 'enemy image' are analysed by Groebel in Chapter 10, which also emphasizes the important strategic role played by the mass media in modern war.

In general it can be said that the main aims of propaganda are to augment love of one's own country in opposition to others. In Chapter 4, Papadakis argued that feelings of glory and love for one's own group cannot be separated from animosity towards others. He further suggested that images of war and sacrifice were central to nationalism. Feshbach (Chapter 11) agrees that patriotism (love of one's country) and nationalism (involving the denigration of others) are correlated, but they are also partially separable issues: some individuals may love their country dearly and be prepared to make sacrifices for it, while showing little tendency to denigrate others, and vice versa. However, both Papadakis and Feshbach would agree that nationalism (as defined by the latter) could lead to the glorification of war. In practice even nationalism is not a military concept: different varieties have flourished at different times and in different countries, and may be differently associated with war (Comaroff and Stern 1993).

In a recent discussion of nationalism, Stern (forthcoming) distinguishes three approaches. The primordialist approach emphasizes attachment to ethnic groups, either by focusing on a primordial need for shared identity or by arguing that ethnicity is an extension of a natural tendency to favour kin (for example Johnson 1986 and 1989). The instrumentalist approach emphasizes the value of nationality and/or ethnicity as organizing constructs for a group. The third 'constructionist' approach emphasizes the socially created nature of nationality. In an important synthesis, Stern assesses the validity of predictions from these approaches. The instrumentalist approach makes a number of non-obvious predictions, but, when taken to imply rational choices by individuals (for example Hardin 1993), fails to account for the power of nationalism when it is counter to individuals' interests, and neglects the emotional effect of nationalism. Stern therefore supports the view (Hinde 1989) that nationalism draws on a primordial sociality which is ultimately a product of natural selection and proximately a consequence of the universal psychological processes involved in group pressure (see above). In this view, altruistic behaviour stems from deep emotional bonds to primacy groups and from socially transmitted rules and norms. These are enhanced and exploited by leaders through the use of national symbols, such as parades (cf. Lane 1981); by tying the idea of the nation to emotionally powerful experiences of individuals (for example suffering); by equating the nation to groups with which the individual already has strong ties (for example the family); and by personalizing the nation. The role of an earlier attachment to the parent in enhancing patriotism, stressed by Feshbach in this volume, is no doubt exploited by the use of terms such as 'fatherland' and 'motherland'.

Earlier (in Chapter 4) Papadakis pointed to the close relations between religion and nationalism. Nationalism can be linked to feelings of permanence and even eternity, and war heroes receive adulation comparable to that often accorded to gods. But religion itself may play a more direct role in war, legitimizing violence and making the risk of death acceptable. Several of the conflicts described in Part I are concerned, at least in large measure, with religious differences. Even some of the tribal conflicts described by G. Lewis (Chapter 2) seem to have had ancestral ghosts lurking in the background, and religion was used to urge soldiers into battle in the Warring States (Chapter 1). A detailed examination of the way in which the Christian concept of sacrifice has been used to glorify war by equating Christ's sacrifice with death in war is given by S. Sykes (1991). Watson (Chapter 12) shows how religious beliefs can operate as a factor that motivates, sustains and legitimizes violence. Fundamentalist conceptions of religion are especially potent. The fighters from the three world religions quoted by Watson draw on religious images and symbols in their explanations of violent action, which 'demonize' the perceived enemy. Murray (Chapter 5), however, sounds a cautionary note on this point: the issues at stake are seldom unitary, and in Northern Ireland the terms Catholic and Protestant conceal differences in economic, political and cultural aspirations (see also Sorabji, Chapter 6). The use of religious labels as a crude model of political difference in Northern Ireland, and the general warlike potential of fundamentalist conceptions of religion described by Watson (Chapter 12), depend on the confusion and conflation of religion and politics-economics. The key point is that the politicized use of religion as a legitimizing ideology can mask underlying social reality and obscure pre-existing patterns of inequality.

There is one issue that is perhaps too easily taken for granted — the differential role of the sexes. This is hardly mentioned in this book, though G. Lewis (Chapter 2) acknowledges the omission. Several others, however, emphasize the importance of manliness or the creation of a macho image in the maintenance of the institution of war (for example M. Lewis and Mosse, Chapters 1 and 9). While at present many of the women who become political leaders may be said to possess masculine characteristics, a basic question is whether this masculine characterization leads to a replication of the macho, conflict-oriented approach in a situation of war. The historical dominance of men in institutional positions of power, albeit with a few notable exceptions, means that the question is largely unanswerable. Undoubtedly, research on the nature and extent of differences between feminine and masculine styles of settling disputes might illuminate critical gendered tendencies in war for compliance, confrontation or compromise. This is not to deny the occasional active involvement of women in war, nor to assert that its infrequent occurrence is totally due to female characteristics and not partly due to the inhibiting influence of men. Nor does it deny a role of women in accentuating nationalism. But given the substantive evidence of history at hand, a greater representation of women in governments could surely only ameliorate the situation.

7. War and Peace: The Socialization of Children

Sharon Smith and Linda Siegel

This chapter investigates some of the factors that contribute to children's socialization towards war and peace. We examine children's home environment, television, toys and the education system as agents that socialize children about war. We also review initiatives related to peace education.

Children's Home Environment

The first members of society from whom children learn about the world are their parents and extended families. What they learn, implicitly and explicitly, helps to determine how they view and act in their world (Brock-Utne 1985). Socialization begins in infancy, and it differs between boys and girls. When parents decorate their children's rooms, gender may dictate the decor. Rheingold and Cook (1975) studied the contents of children's rooms between the ages of one month and six years. Overall the girls' rooms were decorated with lace, frills and ruffles, whereas the boys' rooms had sports or animal motifs. The toys also differed. The girls' toys consisted of female and baby dolls, dolls' houses, and domestic-type toys such as toy stoves, sinks and irons. Playing with these toys gives the girls their first taste of performing traditional domestic chores, reinforcing at an early age that they are to be the nurturers, care-givers and home-makers in their society. The boys' rooms contained military toys such as toy soldiers, construction toys such as cement mixers, and sports equipment. Play with these types of toys involves activities that are directed away from the home, activities such as sports and the military. Therefore, toys can convey powerful messages to the child.

It is not only parents who differentiate between genders but society as well. In a ski contest in Norway in 1983 for eight- to 12-year-old children, the winners received cosmetic kits if they were girls and shotguns if they were boys (Brock-Utne 1985). There are two messages. First, girls and boys play different roles in our society, and second, the role of boys is related to aggression.

Witnessing and engaging in aggressive behaviour is also an important aspect of socialization. How parents interact with each other is as important as how they interact with their children: parents are one of children's earliest role models. If parents are aggressive towards each other, there seems to be a higher likelihood that the children will be more aggressive (Brock-Utne 1985). Children learn from parents what they believe to be acceptable ways of dealing with conflict.

Physical punishment also is linked to aggressive actions in children (Brock-Utne 1985; Watson and Peng 1992). Watson and Peng found that the amount of physical punishment was positively correlated with aggression in children during play. They also showed that boys were punished more often and more severely than girls. Therefore, parents who physically punish their children provide an example of how one uses force to achieve goals.

Girls are not socialized in their home environment to be as aggressive as boys. In general, parents perceive aggressive behaviour in boys as normal and fathers are more concerned with ensuring that their boys do not grow up to be 'feminine' (Brock-Utne 1985). In fact, fathers tend to spend more time, sometimes twice as much, with their sons than with their daughters (Lamb 1977) and are more concerned than mothers about the sex-appropriateness of their children's play (Brock-Utne 1985). Parents, as early role models, teach their children that it is acceptable and appropriate for boys to play more roughly and more competitively than girls and that girls should play more cooperatively. In free play, boys are observed to play more roughly than girls, and to choose toys of a construction or military nature. The girls choose toys such as dolls' houses and play more cooperatively, establishing rules of play. Regardless of the gender differences, if children are taught that aggression is acceptable and no attempt is made to teach both boys and girls cooperative and non-aggressive play, there may be an impact on how they interact with their peers in later years.

Certain types of communities seem to foster more altruistic feelings and actions in children than others (Lieven 1989; Santa-Barbara 1988). Lieven reviews studies that conclude that small rural communities tend to have more altruistic children than large urban centres. Communities that encourage children to be helpful, have extended families living together, have women who are assigned important economic functions, and have positive views on sharing, produce children who are more generous and collectively-minded than communities with different characteristics. For example, children from Israeli kibbutzim performed more cooperatively on a collective task than did children from large urban centres (Shapira and Madsen, as cited in Lieven 1989). Israeli kibbutzim children demonstrated high levels of generosity and cooperation. In this volume, Drucker-Brown has described a Ghanaian people who teach their children that interpersonal violence is to be avoided and that bloodshed is an offence against the earth.

Parents serve as role models also for prosocial behaviour. Rosenhan (as cited in Santa-Barbara 1988) found that adults who were highly committed altruists (for example, non-Jews who risked their lives hiding Jews from the Nazis during the Second World War and white freedom fighters for black civil rights in the USA) came from families whose parents were good role models, especially in the area of altruistic beliefs and behaviours.

Television, Toys and Play

There seems to be an interaction between television violence and children's engagement in war play and other aggressive activities (Costabile et al. 1992; Eysenck and Nias 1978; Watson and Peng 1992). Eysenck and Nias (1978) concluded that there is sufficient evidence to presume that television violence increases viewer aggression. Television is not the only factor that influences violent behaviour but appears to be one of the more powerful ones.

Costabile et al. (1992) asked Italian and British parents about their attitudes towards war play. Both samples reported that a significant number of their children, boys more than girls, engaged in war play. The parents' views towards such play were diverse, although they looked more favourably upon play fighting than playing with weapons or combat figures. The parents believed that war play was popular because of the influence of television and older siblings. The parents felt that the commercials for military/combat toys were very appealing to children. These parents did not support having war toys available in their children's nursery and infant schools. The researchers hypothesized that parents view school as a place where there is 'some core set of values reflecting the "ideal" society' (Costabile et al. 1992: 368).

Watson and Peng (1992) observed three- to five-year-olds to investigate the frequency of aggressive play at daycare and how this correlated with information from a parent questionnaire about children's preferred television shows, toy gun play in the home, children's favourite toys, and the amount of physical parental punishment. Boys demonstrated more aggression, preferred more aggressive toys and television shows, and received more physical punishment from their parents than did the girls. The strongest predictor of physical aggression was the amount of physical punishment the children received from their parents. For boys only, it was found that the greatest predictor of physical aggression was the amount of toy gun play. The level of aggression in the children's preferred television shows was also positively correlated with some types of aggression (pretend aggression where the child used dolls to act out aggressive behaviour or used a non-weapon toy as if it were a weapon) and rough-and-tumble play (play-chasing or play-wrestling). The number of aggressive toys (for example, guns) that the children preferred was positively correlated with the level of physical punishment they received, the level of aggression in their preferred television shows, and their amount of toy gun play (Watson and Peng 1992). These findings, as well as those of similar studies, must be viewed with care. Without a thorough knowledge of what values parents instil in children, the children's peers and the school environment, the picture is incomplete (Ladd 1992). However, the results indicate that aggression is not related to only one factor, but rather children's aggressive behaviour and desensitization to violence may be related to a variety of factors ranging from parenting style to the society's value system.

Television also informs children about what is happening in our world. Surveys have shown that students indicate that television is their primary source for finding out about war. Neither children nor adults always realize that what is depicted on television, even the news, is not absolute reality. Rather, the media depict a constructed world and not reality. News reporters, like everyone else, are influenced by their society. Their perceptions of the world are based on their own value systems. During the Gulf War, the television coverage often presented a one-sided view of the conflict (Goldberg 1991). The news media inadvertently attributed the cause of the war to the supposed personal attributes of Saddam Hussein and the Iraqi army, for example ruthlessness and inhumanity, instead of concentrating on the economic and political issues. The news media did not even attempt to explain how Iraq had become the enemy when the USA had recently been selling weapons to it during the Iran–Iraq war. Hopefully, parents and the education system will help children understand how and why the media deliberately or inadvertently construct reality.

The Education System

The education system plays an important role in children's perceptions of war and peace in a variety of ways. In war each side fights for what it believes to be right. These beliefs are fostered through family, school, peers, community and government/media propaganda (Hinde and Bateson 1989). The combatants want to protect their family, community and values. Before the education system can start to make children understand how to resolve conflicts peacefully, it must fully understand the negative and positive ways in which it influences children. We will discuss some ways in which education socializes children, including social science textbooks and examples of peace education initiatives.

Social Science Textbooks

In general, North American textbooks are written from a very Eurocentric viewpoint, omitting the achievements of visible minorities and women. Textbooks, as well as other aspects of the education system, appear to serve the interests of powerful groups (Anyon 1979).

In a review of 60 US school history textbooks, G. Sykes (1989) concluded that written history depends on the facts chosen by the historian and the facts chosen depend on the historian's own context — 'the cultural and political environment impinges on him/her, affecting his/her perception of the facts and also his/her own understanding of the influences to which he/she is subject' (G. Sykes 1989: 68). Many of the textbooks reviewed were biased and could give the student a false picture of their world.

Sykes claimed that very rarely did a textbook written by an American discuss the Russians in anything but a negative light. Frequently the authors referred to the Russians as 'ruthless and cunning', whereas the USA and her allies were referred to as 'trusting and naive'. In fact only eight of these texts made reference to the fact that in 1918 Soviet territory was invaded by Japan, the USA, France and the UK. Those eight all indicated that the invasion was an attempt to wipe out communism (G. Sykes 1989). Since communism is probably not referred to as a positive form of government, the students, reinforced by their teachers, may make the inference that the invasion was justified. If this were the case, the Russian aggression would be viewed as wrong and US aggression as a necessary response. The inconsistencies in US foreign policy are rarely mentioned in the textbooks — for example, the US support of the Somoza dictatorship in Nicaragua or the US veto of the United Nations Security Council's resolution to condemn the invasion of Angola by South Africa in 1981 (Bigelow 1991).

Recognizing that the US and Canadian education system is different, we conducted a review of some of the social studies textbooks approved for use in Ontario by the Ministry of Education. We present some examples of biases.

In the grade 5 textbook (i.e. for 11-year-olds), *Exploring Canada: Learning from the Past, Looking to the Future*, the authors describe how the native people of Canada and the settlers worked cooperatively in the 1800s for the settlers to learn how to grow new foods and about the best means of travel in the new land (Bowers and Swanson 1985). The authors seem to have ignored the fact that these settlers did not leave a legacy of respect and generosity towards native people. A review of 12 other textbooks made no mention of how negatively the native people perceived the missionaries and Christianity. The text discusses the fact that black people escaped to Canada for 'freedom' from US slave-owners in the 1850s and 1860s, but there is no mention of the fact that Canada had black slaves. Another, but more subtle, distortion is in a chapter on 'Exploring Immigration'. Most of the pictures in this chapter portray the immigrants as visible minorities. However, it does also say (Bowers and Swanson 1985: 294): 'The Inuit and the Indians were the first people to come to Canada. … All other Canadians are immigrants or the descendants of immigrants.'

Another grade 5 text, *Journeys of Exploration*, also provides a very positive picture of camaraderie and cooperation between the native people and European settlers. It also skims over the serious emotional pain inflicted on native people by early missionaries who tried to 'civilize' them by attempting to Christianize them. No mention is made of how intolerant the missionaries were of native religions and cultures.

The senior elementary school textbook, *Japan: Its People and Culture*, puts Japan in the role of an aggressor (Parsons 1988). For example, Parsons (1988: 26) writes that 'Japan saw the nations of the West as threats. The people felt they had to keep up with, or do better than, the West.' This is a very Western perspective. This information can reinforce students' negative views towards Asians.

Canadian textbooks also portray the Russians in a negative light. *Canada's Century*, a high-school text, describes Russia as a country whose 'leaders did not share our beliefs in democracy and human rights' (Evans and Martinello 1988: 300). The authors discuss the Russians as the enemy and a threat. That there are differences between cultures in beliefs about human rights is clear, but it is noticeable that no mention is made of some of Canada's questionable democratic practices, for example, our present treatment of native peoples' demand for the return of some of their land; and our incarceration of Japanese Canadians during the Second World War.

Such biases are not unique to North America. Papadakis (this volume) describes the biases in history textbooks in Cyprus. The Greek Cypriots depict the Turkish Cypriots in a negative light and themselves positively, while the Turkish Cypriots do the reverse, even though both groups are describing the same events.

There are several high-school books that attempt to take a very sensitive approach in presenting history to students. Two examples of these will be mentioned. The first is *Towards Tomorrow: Canada in a Changing World History* (Morton 1988). The author (in the foreword) informs the readers that: 'Historians are fascinated by what happened years ago but they understand those happenings with minds shaped in their own time and they address future readers. However scholarly and objective an historian may be, all knowledge and its interpretation is filtered through an author's experience and prejudices.'

Every chapter asks questions that force the students to think past the facts, questions that challenge the students to think about whether or not certain events and beliefs were fair. For example, there is a picture of a Second World War poster advertising war bonds, depicting a mother and her baby being threatened by a black claw-like hand with a swastika stamped on it. The author then proceeds to ask students to think about what sentiments some of the advertisements used during the war evoked. It also gives an accurate outline of how visible minorities were treated in the late 1800s and early 1900s (for example, the harsh treatment of the Metis, how Chinese immigrants were made to work longer and harder for lower wages, and that when Sikh nationalists came to Vancouver in 1914 they were forced to return to the Punjab region because of protests from Vancouver's citizens); how the Japanese Canadians were treated during the Second World War (for example, how many had their land taken from them and were placed in internment camps); and the attitude towards communists after the war (for example, distrusted, although not to the extent they were in the USA at that time). If the teacher were to explore all the questions asked, the students would gain a much richer knowledge of the past as well as learning about how our intolerance and prejudices can result in aggressive situations.

In another textbook, *Origins: A History of Canada* by Boiteau et al. (1989), each chapter asks the reader to think about the facts. Throughout the text the authors acknowledge that we perceive history from a Eurocentric point of view. For example, they inform the reader (1989: 114) that: 'We say that Jacques Cartier "discovered" the St Lawrence or that Giovanni Caboto [John Cabot] "discovered" Newfoundland. However, we should always bear in mind that we are taking a strictly European point of view.'

The way the text is written forces students really to think about the issues being presented as facts, for example 'Class Debate: Cartier and the French's right to claim the lands of the St. Lawrence for France'. It also attempts to link past events with current issues, for example (1989: 309): 'The question of the right of part of a country to secede is a very complicated one. The issue still exists today. Research one of these areas. Compare your answer with others in the class: (i) Quebec from Canada; (ii) Pakistan from India (1940s); (iii) Palestinian peoples from Israel. Is there any "right" answer to this question of secession?'

Educational Programmes

Bullying is a type of aggression that many children live with on a daily basis. Extensive research has been done on this topic, especially among schoolboys. Studies on Swedish schoolboys concluded that the development of aggressive tendencies in bullies was related to their early and later parent–child relationships, and that the school system can play a significant role in behaviour modification (Olweus 1978). A study conducted at the Toronto Board of Education in Canada concluded that bullying affects children personally from as early as grade 2 (i.e. age eight). Many existing interventions such as the *Peacemaker Program* have helped to decrease the incidence of bullying (Zeigler and Rosenstein-Manner 1991). The school and parents can play an important role in helping children understand why their actions are inappropriate and help them develop more altruistic qualities and tolerance.

Racism is another quality that is learned. Wurzel (1991) interviewed grade 7 students from an urban US school and observed their teacher. The school appeared to be a success in integration. Each class had students from various ethnic backgrounds. But as the researcher paid closer attention he noticed that during recess and lunch the students only played within their ethnic group. When he interviewed the children they all seemed to harbour resentment and prejudice towards members of the other visible minorities. One Hispanic boy (whom the researcher at first mistook for being black) made derogatory remarks about blacks, seemingly unaware that he might be perceived by many to be black. He then went on to say (Wurzel 1991: 32) that: 'Someday I am going to get a gun and I am going to get those *morenos* [blacks].' Other children were adamant about wanting to be considered American rather than being associated with their ethnic group. Not only were these children intolerant of each other, but they were also intolerant of their own ethnic culture and saw violence as a way of solving their problems. It became apparent that the social studies curricula ignored the sentiments of self-hate, inter-ethnic prejudice and violence. The personal realities the children were bringing to school were not being addressed.

Peace Education Initiatives in Two Canadian Provinces

A review of some of the peace education initiatives in Ontario and Quebec indicates that the school system is attempting to address the issues that can affect how a child views the world and chooses to resolve conflicts. Peace education is concerned with the values and motivation that help to promote a broader concept of social conscience beyond one's own community, to decrease aggressive behaviour and to find alternative ways of resolving conflict without harming others (Hinde and Bateson 1989). To accomplish this, peace education must promote knowledge about other people in order to alter our false vision of the enemy. One way to do this is to make children understand that there are many value systems that are different but equal.

It is important for children to realize the benefits of working towards a unified understanding of each other without sacrificing their own identities. For example, a teacher could have the class discuss world issues, such as pollution and depletion of resources, for these universal issues are important irrespective of the cultural goals that differ in other respects. Another example could be to have 'culture sharing' days on which students could bring in items from their own cultures, such as food, dress, or religious symbols. The teacher's role would be to help the students see how much they have in common and how equally valuable their differences are.

During the Gulf War the Toronto Board of Education established some guidelines for teachers in order to help students make sense of the chaos. These guidelines included teachers talking about global conflict with children in the same manner as they would discuss other sensitive topics; listening to the children's concerns about the war; trying to channel the children's emotions into constructive activities such as writing to their member of parliament about their view; and providing information about peacekeeping and peacemaking (Goldberg 1991).

In the USA Wurzel conducted a study at another urban school where he saw how the process of reflective discussion on such issues as the Ku Klux Klan and the Holocaust helped children learn about the connection between history and the human condition (Wurzel 1991).

There is a programme called Facing History and Ourselves (FHAO), which provides resources and workshops that focus on the themes of totalitarianism and democracy and examines issues such as popular resistance, abuse of power and obedience to authority. It addresses hatred, prejudice and racism by examining the collapse of democracy and the rise of the Nazi regime during the Second World War as an example of what occurs after morality breaks down (Goldberg 1991). It can also be used to address contemporary issues in regard to racism. An article by Novogrodsky (1991) gives a poignant presentation of her experience with FHAO. She discusses the positive impact these studies have on students. She recalls one student in particular who was a member of the Skinhead movement (an aggressive racist group who strive to rid their communities of 'non-desirables') before taking her class in FHAO. Although he had been questioning aspects of his group's belief system before he took the course, he credits the course with changing his thinking and behaviour. It made him see that the Skinhead belief system was similar to that of the Nazis and that it was this belief system that had helped the Nazis get cooperation in carrying out atrocities during the Second World War. The student said that he felt the course 'should be compulsory for high school graduation'. This is an example of how racism can lead to membership of an aggressive movement and how education can change individuals.

Another programme in Ontario is the Peacemaker Project at the Downtown Alternative School in Toronto, which is modelled after a similar programme in San Francisco. It helps children learn to settle disputes peacefully (Nobel 1991). Children are trained to work in pairs to help mediate in conflicts that the combatants are unable to solve themselves. The peacemakers' goal is to help the combatants come up with solutions that are seen as fair by all concerned. The programme has been very successful and the peacemakers are admired by their classmates. Setting up a framework for reaching a solution and 'supportive listening' are the main negotiating principles for the peacemakers. The children at the Downtown Alternative School have provided demonstrations of how the Peacemaker Project works at other schools and conferences (Nobel 1991).

Another initiative is an anti-war toys project in Quebec called Pacijou. The project published a curriculum guide for teachers called *Cessez le feu!* ('Hold your fire!') in 1987 (Clanfield 1991). The curriculum was designed to help students become aware that war was not a game and that their world of play was becoming increasingly dominated by toys of destruction and violence. The goal of the project was to show students and parents that there were better ways to settle differences and that there were ways of playing other than with war toys.

The project was divided into three sections. First, elementary school children were asked to give up the war toys they owned so that they could be used to build a public sculpture to celebrate peace. Second, high school students were asked to develop alternative forms of publicity against the purchase and use of war toys and games for children. Third, students were asked to participate in an annual peace march. All the students were encouraged to investigate within themselves and their culture how they had been and were still being conditioned to accept aggression as inevitable and acceptable.

The project was very successful, with 25 Quebec school boards becoming involved; and 12,000 elementary schoolchildren donated over $50,000 worth of their war toys for the sculpture. High school students' anti-war advertisements were featured in 11 magazines and on four radio stations throughout the province. Finally, thousands of students participated in the peace march.

Pacijou is ongoing. In the second phase of the project students will circulate a petition calling for the removal of violent television shows for children and for government to help produce more quality nonviolent shows. This project is attempting to get the school community involved in trying to transform its present culture, which emphasizes competition and often violence, into one of peaceful cooperation (Clanfield 1991).

Conclusion

In this chapter, we have briefly presented some of the many influences that are related to children's knowledge and understanding of war and peace. The complexity of this socialization process must be acknowledged. Education, in both the home and school environments, contributes to children's socialization. Parents, television, toys and, of course, the formal education system, all serve as teachers. All these influences are interrelated. It is important to understand their relationship to each other as well as to the children in order to comprehend with greater clarity the many contributing variables. Obviously, education about peace and understanding from all these sources is critical for the future of the human race.

Note

This chapter was written while L. S. Siegel held a senior research fellowship from the Ontario Mental Health Foundation. L. S. Siegel thanks Marie Thérèse Le Normand for her hospitality during a sabbatical at La Salpetrière Hospital in Paris. The preparation of this chapter was partially supported by a grant to L. S. Siegel from the Natural Sciences and Engineering Research Council of Canada.

8. On the Nature of War and the Nature of Militarism

Hylke Tromp

When Alexander the Great first crossed into Asia Minor, he was offered a suit of armour from the time of the Trojan war, 900 years earlier. He put it on and used it until it became so battered that it had to be replaced (van Creveld 1989: 20). A better demonstration of the lack of change in warfare is hard to imagine.

The conduct of war did not change much for perhaps more than 3000 years. There was no essential difference between the way a battle was fought by the Macedonian Army under Alexander at the Issus, by the Roman Legions commanded by Caesar at Alesia, or by the British armed forces under Henry V at Agincourt. First, the enemy had to be found. Then, battle had to be offered. If the offer was accepted, the armies were deployed in battle order, confronting each other, before the order to attack was given. Sometimes both armies attacked. Sometimes one attacked and the other stayed in a defensive position. Sometimes both sides stayed where they were, waiting for hours in full battle order for the assault of the other side. The impact of the first clash was often regarded as decisive, and horses and even elephants were used to reinforce the effect. Incidentally, Alexander commanded his army in the same way as Henry V did at Agincourt in 1415: fighting in the front line, clearly visible, setting the example. Alexander was wounded nine times during his battles; Henry's helmet, still to be seen in Westminster Abbey, had been badly battered.

Battle, once commenced, was not infrequently a chaotic clash, divided into thousands of hand-to-hand combats in which people tried to capture or kill each other with all means available. It took usually only a couple of hours, and never a full day, before one of the lines broke, which led to chaos, flight and surrender.

The nature of war was determined by the way battles were fought, and their characteristics remained the same during millennia of warfare:

- The armies were always small of necessity: it was impossible to sustain a great army for a long period of time.
- The battlefield was small and easy to survey, and the 'killing zones', where the actual fighting took place, were never more than a few metres wide.
- The decisive weapons were all hand weapons, with most of which it was virtually impossible to wound or to kill at a distance of more than one metre.
- The commander-in-chief was present on the battlefield as one of the combatants, usually leading his army from the front line.
- It was a relatively short clash, never lasting longer than one day.

The Changing Face of War

Changes in the face of battle and, therefore, in the nature of war have come so gradually that defence establishments, as well as security experts, have been able to adjust instead of being forced to rethink. Firearms changed the face of war at first gradually and then suddenly and fundamentally (Dupuy 1980). The introduction of high-velocity rifles and, in 1884, Hiram Maxim's invention of the first perfectly reliable machine-gun, started the age of total war (Ellis 1975). The deadliness of weapons increased suddenly and disastrously. The art of war became the art of practising mass murder in the most efficient way. Johan von Bloch, who between 1890 and 1899 accurately predicted the 'face of war' as it was to be during the First World War, seems to have been one of the very few who realized the changes. The First World War, which started in August 1914, was therefore still officially expected to end with a decisive victory in a decisive battle (as in Waterloo in 1815) before the autumn of 1914. Instead, for four years a continuous slaughter raged at several front lines, the longest 800 kilometres long, stretching from the North Sea to the Swiss Alps. Sixty million soldiers were mobilized. More than eight million of them were killed. On one particular day at the Somme, the British Army lost more casualties than in the Crimean War, the Boer War and the Korean War combined (Middlebrook 1971). The increasing lethality of weapons changed the rules of war and, in so far as the word is appropriate, the culture of war. It forced armies to disperse, and this dispersion led to a steady increase in the size of the armies. This again required conscription and a standing organization: the preparation of war became institutionalized.

During war, the increasing lethality and precision of the firearms forced commanders to direct battle from a safe distance. That, however, meant losing direct contact with the reality of the battle and becoming dependent on incomplete and unverifiable messages, and never being able to survey the battlefield. During the Somme offensive in 1916, the British commander-in-chief was in a castle 80 kilometres behind the trenches. In 1940, Hitler directed his victorious armies from hundreds of kilometres behind the front line. Forty years later, the official commanders-in-chief were supposed to keep political command and control over distances to be measured in tens of thousands of kilometres.

In the meantime, war had become nothing more than slaughtering soldiers from a safe distance. When this failed to produce victory, civilians too became targeted for annihilation. It took more than a century, two world wars and the invention of the ultimate weapon, the atomic bomb, before the impact of this change started to become fully realized: war had become 'total war'. Warfare in the twentieth century is now an industry. It is bureaucratized, to the extent that its main decisions are being taken anonymously and committed to paper by people far removed from the actual killing zones.

After the Great War (1914–18), the conclusion was (temporarily) drawn that war had to be abolished as a conflict strategy, as an official and legitimate way of settling conflicts between states, as a continuation of a state's policies by other means. But despite official declarations, war preparation and in particular the threat of war continued to be seen as necessary and even indispensable: '*si vis pacem, para bellum*'. It became official policy again to be prepared for war at all times and under all circumstances. At the end of the cold war, the highly advanced, steadily modernized new means of mass destruction were deployed in such a way that there was not one target on the globe that could not be destroyed within minutes; and the whole world and even outer space had become potential battlefields. All legal and formal distinctions between soldiers and civilians had been superseded by factual developments: the weapons were not able to discriminate between them, so both were targeted.

There are different ways to summarize the changes in the nature of warfare during the last century, but 'loss of control' seems to be the most appropriate description. Wars have always had unpredictable starts and unpredictable outcomes; but in the twentieth century the outcomes have always been far beyond the imagination of the decision-makers who were responsible for their beginnings. The First World

War caused the deaths of 14 million people and led to social upheavals, financial disasters, economic crises and revolutions; but it did not achieve any of the limited goals the decision-makers had in mind when they went to war in 1914, and it destroyed most of them. The German, the Austro-Hungarian, the Russian and the Ottoman empires dissolved in turmoil as an immediate result of that war, which was followed by communist and National Socialist totalitarian dictatorships in Russia and Germany and, finally, by a second world war. The Second World War killed approximately 55 million people, destroyed a great deal of Europe, and ended with nuclear bombs annihilating Hiroshima and Nagasaki. It, too, changed power in unforeseen and unpredictable ways: it made the United States and the Soviet Union hegemonic world powers, and it divided Europe into two parts. It led to the end of the British, French and Dutch colonial empires, all consequences once again way beyond the imaginations of the people who had made the decision to go to war. War had become an even more dangerous adventure, more resembling Russian roulette than Chinese chess. Now, even more than before, its outcome is unpredictable and tends always to extend far beyond the limited purposes of the politicians who declare war.

Warfare at the End of the Twentieth Century: The End of Yugoslavia

War came slowly to Yugoslavia. There was hesitancy on both sides. But uncontrollable forces provoked commitments, rumours of atrocities mobilized fears, and then it became clear that there was a successful strategy to conquer territory. Serbian irregulars started what later would be called 'ethnic cleansing'. Surviving refugees have described how it worked. They got a warning that their village would be shelled; often the warning was no more than the sudden disappearance of their Serbian neighbours. This was the sign that they had to flee. If they did not, the village would be shelled from a distance until everybody had left. Next, Serbian militias and irregulars would occupy the village. Any remaining resistance would be rounded up: during the civil war in Croatia this was usually with the help of Federal Army forces.[1] Those taken prisoner were killed or deported to a concentration camp. The irregulars did not belong to the Federal Army, but their weapons came from its arsenals and from the depots for territorial defence.[2]

Gradually the war not only increased in ferocity, but intensified in another sense as well. First there were rumours. Then concentration camps and mass graves were discovered. Survivors as well as murderers told their stories to the mass media.[3] Their stories revealed that the atrocities of the civil war between 1941 and 1945, during the Second World War, were being repeated. This civil war, afterwards covered up in official history behind proud stories about resisting German occupation, had been a murderous civil war in the first place, in which more than one million Yugoslavs were killed. In 1992 'ethnic cleansing' was officially declared to be official policy: a sharp reminder of the devastating Balkan wars in 1912 and 1913, later described as fought for 'the complete extermination of an alien population' (*International Herald Tribune*, 24 November 1993).

Medieval warfare 'does appear to have been dominated by two general principles: fear of a pitched battle, and a "siege mentality". … War is … above all made up of pillaging, often of sieges, sometimes of battles' (Contamine 1984). The fourth Balkan war this century is a medieval war fought with modern weapons: means of mass destruction, designed to kill from a distance. It is a medieval war only in strategy; it is a modern war in the use of modern weapons. Cities and villages are shelled from a safe distance. Individual people — unarmed civilians — are targeted and killed by snipers. When a siege ends in occupation, the conquering forces do what medieval armies did hundreds of years ago; they pillage, rape, kill, burn down and destroy. There is no record of a 'pitched battle' between regular forces. When 'fierce fighting' is mentioned in press reports, it turns out to be an unexpected clash — for example, when Serbian militias enter a village or city on the mistaken assumption that it has been abandoned due to a long bombardment. Since only the Bosnian Serb forces have the heavy weapons of the Federal Army, its opponents, the Bosnian Croats and Bosnian Muslims, have no choice other than to wait, hiding in shelters until the Serbian forces consider it safe to occupy the village or city. Then they can open fire. For that reason, the siege of Sarajevo has already lasted for more than a year and seems unlikely to end: the defenders have no possibility of escape. They, too, know about concentration camps and mass executions.

The war in Yugoslavia has now escalated into several wars simultaneously. It has the characteristics of an international war, fought between official Serbian, Croatian and Muslim armies. At the same time, it is a war of independence, fought by irregular militias. It is a civil war in which, as in all civil wars, the laws of war are completely ignored. It is a civil war like the Spanish Civil War in which the regular army (the Yugoslav People's Army, now predominantly Serbian) is the dominant force on one side. It is a civil war like the Lebanese civil war because there are many different irregular groups and militias, killing each other for power, money or just revenge. It is a civil war like the civil war in Somalia, fought by warlords and their mobs.

Moreover, it is a war which continually threatens to escalate outside the Yugoslav borders as it involves minorities of its neighbouring countries, in particular in Kosovo (Albanians) and Vojvodina (Hungarians), while the Muslim population in Yugoslavia looks for help to Turkey and the Arab World. Finally, since 'peaceful nuclear explosions' are offered on the free market as a result of the collapse of Soviet empire, this civil war might even become a nuclear war: the regular threats are to be taken seriously.

In the Middle Ages, winter usually ended the war season. Hostilities were resumed in summer and autumn, since fighting during the winter was simply impossible. Modern war, however, is of all seasons. The sieges continue and the intensity of the killing even increases. There is no end to this kind of war. Those defeated in the first round will try to reconquer in the second. Such a war will escalate, and then, unexpectedly, it will seem to have come to a logical end when one of the parties appears to have been defeated. But then, the war will erupt again unexpectedly, to be ended once more by general exhaustion. It is 'a state of war' rather than a war, in which long periods of relative tranquillity will be regularly interrupted by outbursts of violence. In the late Middle Ages, there was a Hundred Years' War between England and France (1337–1453). Later, there was an Eighty Years' War between Holland and Spain (1568–1648), and a Thirty Years' War in Germany (1618–48), which killed almost half the population. More recently, France and Germany fought major wars in 1870–71, 1914–18 and 1939–45: during the violent interludes in a period of almost a century they considered each other as arch-enemies (*Erbfeinde*). Whenever during this period there was no actual fighting, preparations for war were under way. An equally stable 'state of war' now characterizes the Middle East. 'Stable Peace' — no threat, no expectation, no preparation for war — is not easily achieved, but it exists; and therefore, it is possible (Boulding 1979).

On Militarism and the Nature of War

On 27 November 1992, the *New York Times* as well as a great number of other newspapers published a story entitled 'A Killer's Tale — A Special Report: A Serbian Fighter's Path of Brutality'.

> What Borislav Herak remembers most vividly about the sunny morning in late June when he and two companions gunned down 10 members of a Muslim family is the small girl, about 10 years old, who tried to hide behind her grandmother as the three Serbian nationalist soldiers opened fire from a distance of about 10 paces. … Mr Herak also recounted being taken to a small farm outside Vocosca where a 65-old-Serbian volunteer … had led Mr Herak and three other young Serbs to a grassy plot one morning in early June and shown them how to wrestle pigs to the ground, hold their heads back with their ears and cut their throats. Days later, Mr Herak said, he used what he described as 'this skill' to cut the throats of three Muslim men captured fighting for the Bosnian Army.

Borislav Herak was at the time 21 years old — 'a primary school graduate who pushed a handcart for a living at a Sarajevo textile company'. He was 'drafted' into the 'Serbian Volunteers'. His father is a Serb, his mother was Croat; his sister is married to a Muslim, whom he holds in high regard. Mr Herak said he had had nothing but good relations with Muslim neighbours. But, from Serbian radio and television and in gatherings with other Serbian fighters, particularly the older generation steeped in Serbian folklore going back to defeats by the Ottoman Turks in the Middle Ages, he said he learned that Muslims posed a threat to Serbs. 'We were told that we would have to cleanse our whole population of Muslims,' he said. 'That's what we have been told. That's why it has been necessary to do all this.' However, the young fighter said that he also had been motivated by the urge to have things he never had before the war, including women and items like television sets and videos and foreign currency. Mr Herak described how he and a companion had attacked (i.e. raped) Fatima, whom he described as a nice woman, about 30 years old, in a room at the motel and then taken her at gunpoint in their car to Zuc Mountain. 'We stopped by a small bridge, and I told her to get out. She walked about three metres away from the car with her back to me and I just shot her, I think in the upper back of the head.'

On 26 October 1993, the portrait of another 'fighter' was published on the front page of the *Guardian*. Caption: 'The face of the "amiable lad" who devastated the Shankill Road … Thomas Begley, the IRA assassin who killed ten people including himself. "He had nothing against Protestants," friends said yesterday.'

From the accompanying article, it appears that the man who carried a bag of Semtex into a fish and chip shop in Shankill Road with the intention of blowing up the UFF leadership, but instead killed nine innocent civilians as well as himself, was an unemployed 23-year-old: 'a very quiet bloke ... who didn't drink much ... loved playing cards, was near enough a mechanic, loved working on cars. ... He had nothing whatsoever against Protestants. ... He worked alongside them building roads in Bangor and when they finished they all went off for drinks together.'

Both 'fighters' were just 20 years old, unemployed and not very well educated. Neither had anything personal against the people they killed. They believed the stories they were told and did what they were ordered to do.

In such cases, soldiers no longer fight in the sense in which that has always been understood. Fighting ('combat') has been replaced by murder from a safe distance, often of people who have no weapons and cannot harm anybody at all. In official war plans, area grids on maps are targeted to be destroyed by artillery or high-altitude bombardments, an activity that bears considerable resemblance to the genocide in concentration camps. In terrorist strategies — often depicted as the poor man's military strategies — individual persons or small groups are targeted to be destroyed by snipers' bullets or bomb attacks. The scale is different. The strategy is the same. The consequences are identical too: the victims may have nothing to do with the conflict and may be innocent bystanders. In all strategies, the killing is done from a safe distance. There is no possibility for defence, let alone a fighting chance.

The behaviour of the 'soldiers' remains determined by the same factors that have always been present in the history of warfare. Why do they accept and obey orders that are in sharp contradiction of the values and norms into which they have been educated? What is the 'combat motivation' and what determines 'combat behaviour'?

Alienation

In normal life, everybody is part of quite a lot of groups. Individual behaviour is shaped by the expectations or even prescriptions of these 'reference groups'. Everybody's behaviour must conform continually to the expectations of relevant but different reference groups. As a consequence, participation in social life consists of playing different roles, depending on time and circumstances. One is paterfamilias or child, driver or pedestrian, teacher or student, blue-collar worker or so-called intellectual, Christian or Muslim, soldier or civilian; if soldier, other subdivisions become relevant such as from army, navy or air force, regiment and rank. Roles and therefore behaviour change if group participation changes. In general, the 'behavioural expectations' of one's reference groups are not contradictory. They are based on a generally accepted set of values, norms and even laws: a culture (Dahrendorf 1959).

However, once recruited into army life, a soldier becomes increasingly isolated from his previous and all other reference groups, even from his family; and during a war and in the fron tline, the alienation from civilian life is complete. He is forced to adapt to the culture of war and to adapt to or accept a completely different set of values and norms.

This alienation from normal life explains, among other things, war crimes. Norms in civil society are not only different, but in complete contradiction to the norms of war; and soldiers have to invent these norms in order to survive. They regularly confront life-threatening situations not foreseen in the army's handbooks — warfare always escapes the handbooks, guidelines and predictions, which are based on experiences of former wars. So they develop their own answers to these situations and it becomes the norm to minimize risk: shoot first, ask questions later. Especially in guerrilla warfare, there usually is at least one party who cannot be expected to follow Geneva rules; and the inclination to kill any person who might be dangerous tends to become a safety prescription and eventually a norm: a part of the 'culture of war' (Marshall 1967; Moskos 1970; Ellis 1980, Holmes 1985). This culture of war can include committing horrifying acts of mutual revenge, since innocent and harmless civilians may be killed first as a precautionary measure and then in retaliation, usually justified in terms of deterrence, until mutual killing of unarmed and defenceless civilians, as well as burning down and destroying their houses and villages, has become a 'normal' act of war: normal in the culture of war, which has replaced all values and norms of civil society. Therefore, it is not surprising that virtually all armies during war have been known to commit war crimes. The mass murders of Croats by Serbs in Vukovar in 1992, or the mass murders of Muslims by Croats in Ahvinici in 1993 are not the symptoms of a specific Balkan culture — neither are the mass murders by Germans in Lidice or by Americans in My Lei symptomatic of the degree of civilization in Germany or the United States.[4] The Dutch army in the Dutch East Indies, the French army in Algeria, the British army in all their colonies have all committed the same kinds of atrocities, now usually called 'war crimes'. It is the culture of war that prescribes such killings, and committing 'war crimes' is therefore inevitable.

Social Control

This was a dominant factor in explaining combat behaviour so long as soldiers were fighting in ranks close together and all behaviour was clearly visible. It diminished in importance when the ranks were broken and when trenches and manholes hid the soldiers from the enemy and from each other. Participation in warfare diminished accordingly — to only 15 per cent of the whole force according to some studies. So long as behaviour is visible, social control is possible. Soldiers know that their behaviour is seen and judged by others, and they would rather not be seen as cowards.

Discipline

Implementing disciplined behaviour has always been the foundation of all army training. Much of the soldiers' behaviour is programmed: combat behaviour should become completely automatic, without any interference of emotions or cognitions, almost as a Pavlovian conditioned reflex. Killing should become an instrumental act in which morals, values, norms and laws cannot interfere when the military command is given by the appropriate authority, who is easily recognizable by rank. The authoritarian structure of all armies aims at the dispensation of individual responsibility: '*Befehl ist Befehl*'. The properly trained person knows that not he, but the person in command, is responsible for what he does; and what he does in war is kept separated from his normal set of values and norms.

Ideology

Ideology (defined here as a more or less coherent set of attitudes and beliefs with a propensity to act) plays a role, but, contrary to what is generally assumed, it is often a minor one. Soldiers are not motivated to do what they do by any of the high ideals for which they are assumed to be fighting. High ideals, even nationalism or patriotism, have run their course by the time the battle starts. The sobering assessment is that ideologies sometimes motivate people to enlist in an army, but that thereafter they soon disappear as motivating forces. It is unusual to find a veteran of any war proclaiming that patriotism or the defence of democracy or another high ideal had motivated him to do whatever he did; and since modern war is nothing other than committing criminal acts on a major scale, this is not something to regret. It should be noted that it is not unusual for soldiers to change sides during a war. Alexander simply took over command of Darius's armies when the Persian emperor was defeated and had been murdered by his generals. Thousands of years later in the Second World War, one million Soviet soldiers were recruited by the German High Command to fight against their own country. The Yugoslav People's Army has for 40 years been trained to defend its country against any foreign intruder; now its soldiers are destroying it in a way no foreign power would ever have done. Ideology passes away as a motivating factor; its main role is probably to keep up appearances of respectability and justification afterwards, a process usually described as rationalization in psychological jargon. As a moving force, ideology, any ideology, is easily expendable.[5]

Borislav Herak and Thomas Begley represent the two extreme types. Both murdered people who could not defend themselves, as is usual in such wars. But there seems to be a major difference in what motivated them to act the way they did.

Herak was recruited, indoctrinated and subsequently completely isolated from 'civil society' and its values, norms and culture; in contrast, Begley was not. Herak was told what to do by people whose authority he had to accept; he followed orders and, in case of doubt, interpreted them in the same way others did; he therefore killed the women he had raped. In the end, he escaped or was accidentally taken prisoner while trying to get out (it is unclear which) and immediately started to tell his story. His situation was no different from that of the American conscripts of the same age who, during the war in Vietnam, murdered women and children in My Lei (and other places): they were alienated from normal life, under strong social control, subjected to peer pressure, officially relieved of all responsibility for what they were doing and educated into justifying killing, any killing, as being for the good of the country or for freedom and democracy.

Thomas Begley, who was not drafted into any army, lived in a relatively quiet, normal civil society (the average number of killings in Northern Ireland in any given period would, in many American cities, be considered negligible). Still, his behaviour can be explained and understood within the same framework. Alienation from the normal culture does not necessarily have to coincide with geographical isolation, as in war; it is a psychological process, which can be enforced by others, but can also result from a conscious decision. The phenomenon is symptomatic of members of terrorist groups in Western Europe, who hijack, blackmail and kill to destroy the society in which they live, and may even perform normal roles to hide their real purposes and actions.

Here, too, the authoritarian structure characteristic of all armies plays an important role. It appears, and not only in this case, that the youngest, most susceptible people are always drafted for the most dangerous missions by people they respect and trust because they are in command. Thomas Begley's face is surprisingly like the faces of the soldiers killed in the Great War; 75 years ago he would probably just as easily have been recruited to perform a suicide mission, or to go over the top near the Somme or Ypres. Moreover, social control is at its strongest in small groups: in front of others it is almost impossible to refuse a task that in the context of the prevailing ideology is depicted as necessary, honourable and heroic.

Between Borislav Herak and Thomas Begley stand the majority of the modern soldiers who commit murder from a distance and never from nearby; or are accomplices in performing these tasks. The final act of almost all military action is pushing a button; in its most modern version, the consequence is the extermination of an area tens, hundreds or thousands of kilometres away and watched on 'real time' television via satellite coverage. The alienation from reality is total: human suffering or disintegrating and dead bodies will not be seen.

For millennia, war has been equated with fighting and winning battles; and militarism in its most positive meaning emphasized the relevant virtues, such as courage, competence and discipline, as well as honour and honesty. There were rules of war, which was once described as the game of kings; and the rules of the game had to be honoured. And there have been times, when war was, at least to the high and mighty in medieval Europe, indeed a game, and tournaments as well as wars were fought by teams of knights according to preset rules, resembling the current European soccer competition in more than one respect: the battles ended with many of them severely hurt and even wounded, but not dead, since that would not have been according to the rules, and not profitable either.

Changes in the nature of warfare have also changed the nature of militarism. Militarism is now seen as as an ideology closely linked to authoritarianism and dogmatism. It is characterized by, among other things, rigidity, inflexibility, political conservatism and, increasingly, a lack of intelligence or political education. As an ideology, it is at one end of a scale on which 'pacifism' is at the other. This new 'militarism' is therefore regarded as the psychological foundation for murderous and criminal behaviour during war. It is not. The reasons why people kill and murder are different, as I have tried to make clear in this contribution.

Notes

1 According to General Blagoje Adzic, who, on 8 May 1992, resigned as chief of staff of the Federal Army, the war in Croatia was inevitable because it was prepared, planned and imposed. 'In Croatia, we had to defend the Serbian People from Fascist-Ustasja genocide. In the beginning, the Army was preventing that genocide by dividing the conflicting parties, and then the Army was directly attacked.' Interview in NIN, 5 June 1992: 31–3.

2. It is interesting to note that during the cold war, the Yugoslav defence system was by many in the West advocated as setting an example for what was advertised as a 'non-offensive defence' system. Another example was Switzerland; and the main thrust of the argument was that the best defence would be created if the whole population had access to arms to defend the country against foreign aggression. In Yugoslavia, they had. This is one of the main reasons why the killing continues.

3. The most famous case is that of the Serbian soldier Herak, who was extensively interviewed and whose confessions were repeatedly shown on television; quotations from his story are included in this contribution. He confessed to having committed a number of war crimes, but what is relevant here is the part where he explained how he was forced to do it, and that it became clear that he was only one of many. His whereabouts are presently unknown since he has been convicted by a Croatian court of committing the war crimes to which he confessed.

4. These and other empirical investigations have revealed that during combat situations most of the soldiers tend not to fire their weapons: they rather remain invisible, taking literally a low profile. It may now even be assumed that an extremely high percentage of the soldiers sent to a front line — the 'killing zone' — tries to escape participation; which, incidentally, is both a consequence and a cause of the increasing reliance on the use of firepower from a distance. One obvious way to escape war is desertion; another is mutiny. Mutinies during war are usually suppressed or neglected in official histories. But in 1917 almost the whole of the French army refused to fight. In 1918 Germany collapsed, in some measure due to a mutiny. During the Vietnam War, officers of the US Army were sometimes killed by their subordinates ('fragging'). Desertion has been even more common. During the Middle Ages, armies were in constant danger of dissolving by desertion. During the First World War, many soldiers probably deserted, some trying to keep alive even in no-man's-land or in other 'sanctuaries'. In 1944, on the

eve of Germany's last offensive in the Ardennes, Hitler's adjutant Martin Bormann concluded that more then 400,000 German soldiers were AWOL (away without order of leave: actually, had deserted): a number twice as high as assembled for the Battle of the Bulge.

5. Not all authorities would agree with this view. Ideology was certainly not absent in the Second World War and may have played an important role in motivating the German army on the Russian front.

9. The Knights of the Sky and the Myth of the War Experience

George Mosse

The myth of the war experience is crucial to the institutionalization of war: it masks the reality of war and cleanses its memory. The myth that surrounded the First World War sought to transform a negative into a positive experience through linking death in war to the glory of the nation and through encompassing the hopes and yearnings of that generation. The 'conquest of the skies' was one important way in which the reality of war was masked and made bearable, though its implications reached much further, to a way of looking at the nation and the world which, in turn, made confrontation with modern war that much easier. Wartime aviation presented a climax of the myth of the war experience.

It seems almost incomprehensible in hindsight that tens of thousands of young men could have volunteered for war in 1914 filled with enthusiasm and hope. To be sure, they did not know what awaited them, but the expectations that carried them to war were also those that helped create the myths that surrounded the war in the air. The quest for action, for the exceptional, motivated that generation, an escape from a middle-class life that seemed to constrict them. There was nothing unusual about that. Already when the first volunteers in history had gone to war, 100 years earlier, the German poet Friedrich von Schiller had written (in 1797) in a much-cited poem, the so-called song of the cavalry (*Reiterlied*), that the soldier who discards the anxieties of daily life and has no steady home alone is free — for he faces his fate boldly and looks death in the eye. And now, in 1914, in one of the most famous English war poems, somewhat ironically called *Peace*, Rupert Brooke talked about discarding the littleness of life. To him volunteers in war were like swimmers diving into cleaniness, leaping from a world grown old and dreary. Certainly this feeling was widespread and it implied several other blessings: the war as a test of manliness, a quest that would carry over into the postwar world, a longing for purity — moral and sexual purity — in opposition to what some of them saw as the general degeneration of a world grown old. And such sentiments were accompanied by the consciousness of the energy and joyfulness of youth.

The quest for purity, the test of manliness and the thirst for action were all interrelated, the product both of education and of rebellion — an education which, on the one hand, had stressed so-called manly qualities of willpower, courage and patriotic service and, on the other, rebellion against what seemed a passive and fossilized, or better, routinized society. To be sure, we now distil these attitudes from a handful of European writers, but it was a minority of articulate intellectuals who set the tone, created a lasting myth of the war experience, and here the volunteers played a leading role.

Yet this enthusiasm soon cooled in the trenches, the reality of war determining youthful lives. The wish 'to fight for Germany true and noble, free from deceit and wickedness' gave way to loss of life, mud and boredom. The battle of the Somme in 1916 seems to have crowned a long process of relative disillusionment, with its million dead and its inconclusive outcome. But the word 'relative' is important here, for the need to mask the war was all the greater, to make the hourly risking of life meaningful in some manner that would make it possible to confront the horror and still see some of the glory of war as they had understood it.

The ideals, wishes and hopes of 1914 were now largely projected from the earth to the sky, those who scaled greater heights becoming the focus for those expectations of adventure, liberation and manliness, which fared so poorly in the trenches. Benito Mussolini, then a corporal, wrote to a pilot during the war to say how much he envied him in contrast to his own enforced idleness in the trenches. Mussolini himself became a passionate aviator. We know about the reaction of front line soldiers who craned their necks to watch the battle in the sky, who felt resentment and at the same time craved for the restoration of perspective on war. The 'conquest of the skies' seemed the kind of battle that the war should have been, but was not — here was the freedom, the purity, the demand for individual courage and power of will, which presented such a contrast to the almost impersonal war on the ground, the battle of machine-guns and grenades. Flyers were the 'knights of the sky', bringing together infinite vistas and medieval metaphor, the one most frequently used to mask war. For example, substituting a sword for a gun was common in war postcards all over Europe — especially those showing children as soldiers — but also on war monuments after the war. Chivalry denoted individualism for both friend and foe, a method of fighting that seemed completely different from the reality of modern war. But it also meant passing a test, a true test of manliness with all that this implied. There is a continuity in the traditional stereotype of masculinity here quite unconsciously expressed by Samuel Hynes in his memoir of a combat pilot in the Second World War: 'and because I hated it, I did it; you could only evade what you were not afraid of — that was understood' — what an unthinkable remark to make in the trenches, or indeed during the war on the ground.

Chivalry and manliness, the wish for liberation and patriotism as well, were part of the self-consciousness of youth. The word 'young' was a prefix to every heroic deed in the war, to all men marching into battle; it meets us on all sides. The volunteers of 1914 stressed their youth in order to set it off against the society of their elders, and enlisted the war to that end. The war, so one of Rupert Brooke's famous war poems had it, 'caught our youth and awakened us from sleeping'. Many of that generation, whether rebels or patriots, grieved for their prematurely lost youth and searched for the unusual, the extraordinary; they found it in war, just as they thought they would find in war the manly virtues that had never before been tested. But this worship of youth also became widespread in society as a whole during the war. However, there were also some middle-aged men who in this way wanted to recapture their youth, such as the 50-year-old German writer, Hermann Löns, who volunteered and was killed at the front.

Youth was an ideal the flyers exemplified in their own persons. They were, in fact, very young as a group, while many of the conscripts in the infantry were young adults or indeed middle-aged. By their very youth they cut heroic figures, but they also exemplified the myth of youth as it should be: through their courage to the point of recklessness, joy in combat, and disregard of death and danger. Here was an exemplary way of coping with war, so it seemed, distilling its adventure, exaltation and vitality, and sublimating its horrors. Once again, those who conquered the skies exemplified the hopes of those who had rushed to the colours and now found those hopes dashed by reality.

But myth must have some basis in reality. Or did fighter pilots themselves come to believe in their myth? Pilots did treat their enemy with some chivalry: for example, they would drop wreaths behind enemy lines to salute a fallen foe. Manfred von Richthofen, the German ace, was given a solemn funeral by those who shot him down, who then dropped a metal box with pictures of the actual funeral behind the German lines. Von Richthofen's own squadron entertained English flyers who had been shot down and survived. Certainly, until 1918 the battle in the skies was largely one of single combat, and even in the Second World War Samuel Hynes thought of himself as alone in the sky, for he could see neither his tail-gunner nor his radio operator. The existence of 'knights of the sky' seemed proof that war could still be a chivalric adventure. Italo Balbo, fascist chief and passionate aviator, summed up its implications very well when he said that 'through aviation Italy has rediscovered the chivalry of the past'. A clear-cut victory was still possible, as planes came tumbling out of the skies. Here was proof that daring counted, a manly quality which had formed part of the education of that generation, but which in the trenches meant a sure death.

Nevertheless, reality intrudes here as well. As Eddie Rickenbaker put it: 'Fighting in the air is not a sport, it is scientific murder.' Aerial attacks are a matter of 'military-technical precision' — no medieval image here. A member of Manfred von Richthofen's squadron even called himself a 'machine-man'. Side by side with that of chivalry stood a modern, different image of the pilot. These were no longer the 'last of the knights' as a German book on wartime aviators published in 1931 has it — those who in contrast to the other armed forces were dependent on themselves alone — but instead here 'machine fought machine'.

This view of flying, though nearer to the truth of actual combat, also stripped the horror from war by depersonalizing it, taking, as it were, the human factor out of the battle. But this construction of a myth of the war experience was apparently flawed, for despite the man-machine even here the ideal of the pilot as exceptional, dependent on himself alone, persists, if in a more realistic manner than in the dominant image of knighthood. The same book that told of machine fighting machine goes on to say that nevertheless this was no butchery but a knightly combat, that pilots have their own unwritten laws which they cannot share with men on the ground. Ernst Jünger, the German popularizer of war, for example, saw wartime aviators as a commanding breed, a race that built machines and trusted machines. This new race was hard and pitiless, possessed of technical competence, but also of those characteristics common to the individual manly ideal — 'iron nerves, steady eyes and quick decisiveness'. These qualities, however, were not confined to flyers; they inform the famous German poster calling for war loans by Fritz Erler, or Ernst Jünger's characteristic of the front line soldier as him 'whose eye has seen a thousand deaths', part of a 'new race of men'. Flyers could live this manliness in their kind of combat, demonstrate it in action for all to see. They exemplified indifference in the face of death, which was the supreme test of manhood in war. Contemporary biographies of the flying aces usually stressed such indifference, just as the myth of the war experience itself can be interpreted as trying to erase death as the bottom line of war.

Such manliness, as well as the idea of chivalry, transformed wartime pilots into a new elite: new, at the time, because not aristocratic by background (flyers came from all social classes). All flyers were officers who were apt to receive high military decorations faster and more frequently than infantry, and in Germany and Italy many of them were part of a military elite in the first place: they had already been preselected as pioneer mountain troops or cavalrymen. Pilots had good reason to look upon themselves as an elite, just as those on the ground saw in them a myth about war come alive.

But the myth of the 'knights of the sky' had a still more general significance, for it did not stand alone as a climax of the myth of the war experience. Aviators were not the only ones who conquered the skies in quest of purity, far removed from the squalid reality of war. Mountain troops considered themselves an elite as well, for mountains were the '*fonte purissima de spiritualità*', as an Italian tract on mountaineering tells us. Mountain troops also had a tight-knit camaraderie, and just as significant, as one of their historians has it, the real challenge was nature rather than the enemy. This was true for flyers as well — and nature had much the same significance: the elements were often hostile, but above all military action seemed to take place within timelessness. The constraints of time fell away — this was a real disengagement from the dreariness of daily life, for which so many of the volunteers had longed. In the midst of the Second World War, Adolf Hitler, who had himself volunteered in the First World War, remarked that he had built a house in the mountains not because of the beauty of the landscape, but because 'here I am far from the littleness of life'. These were the sentiments of Rupert Brooke redeemed through war. Mountain climbers and mountain troops still felt what Edward Whymper, who had been the first to climb the north wall of the Matterhorn, had written in 1865, namely that his was an act of physical and moral regeneration.

Mountains had always fulfilled sacred functions, there is hardly a European nation without its sacred mountain, and now the 'knights of the sky' were added to mountain magic as symbols of moral regeneration. The quest for purity has been mentioned earlier and here it also took on symbolic meaning. Conquering the mountain symbolized innocence and integrity: as Louis Trenker, one of the chief German popularizers of mountain magic said shortly after the war: evil people do not climb mountains. Such sentiments, which were applied to courageous climbers or to individual flyers, were co-opted by the nation for its own ends: it could make good use of this myth for purposes of self-representation. The nation that presented itself as eternal, that wanted to pull down a piece of eternity into people's lives and make the rush of time stand still, had obvious affinities with the sky and the mountain. Purity was associated with a specific nation or a specific race, with an ideal type or with the search for a national character. The so-called 'outsider' was different: impure, unable to scale these heights. In all nineteenth-century German literature, no Jew ever climbed a mountain.

Such nationalism was not a necessary part of the myth of the 'conquest of the skies' in the First World War, or rather, patriotism was taken for granted, while the ideal of chivalry, as we have seen, did lead to some respect for the enemy backed up by the actual conditions of aerial combat. For example, it was said that Oswald Boelcke, perhaps Germany's most famous air ace during the war, would never kill an unarmed enemy. Even on the ground, astonishing though it may seem, hatred of the enemy was only occasionally used as a means of coping with life in the trenches. The conquest of the skies led to thoughts of earthly domination only after the war.

The air ace then symbolized the 'new man' or 'ideal type'. Through his comportment and looks he exuded scorn of danger and control over his body and nerves, that spiritual chasteness that knew no evil. He focused, more than any group could have done, the hopes that had fuelled the generation of 1914 — and he had conquered! The pilots were not only members of an elite, but they also furnished heroes in a war filled with acts of individual heroism, but none so spectacular as those of the 'aces' Manfred von Richthofen and Eddie Rickenbaker, whose names are still remembered when those of the generals of that war are mostly (and rightfully) forgotten. Wartime propaganda in the search for heroes supported this image, which could set an example. The phrases 'hard as steel' and 'steely blue eyes' were often associated with such heroes, a manliness which had, in fact, been an ideal in

European society long before it proved its true worth in war. This was the case in all the warring nations, though Germany in particular could use such aces to demonstrate that the material advantage of the allies could be overcome, emphasizing the individual in a war in which the machine had become dominant. The 'conquest of the skies' was basically a myth directed against the times. It combined man and machine in one image which, through the virtues discussed above, served to personalize war and thus to disguise its real nature as it was fought on the ground.

Flyers had the advantage here: there were no 'aces' among mountaineers, at least not in the war. For all the mountain magic this was a war on the ground and it was, for example, the 'Alpini' as a whole that postwar Italian literature celebrated. However, in defeated Germany after (rather than during) the war, mountain magic did produce its own aces. The vogue in the early 1920s for so-called mountain films with Louis Trenker and Leni Riefenstahl used the beauty of nature, the wide open spaces (glaciers were prominent) to make present-day reality seem puny and unimportant. But perhaps, just as significant, youth could find here what it missed in peacetime: men fighting battles in the midst of constant danger — as Louis Trenker put it — struggling in close proximity to death, leading to heroic deeds and hard-fought victories. Thus Trenker in 1931 once more appealed to the 'Spirit of 1914', and not quite in vain.

The mountains, like the skies, were used to mask the truth about war and supposedly helped men face mass death with greater stoicism. The 'conquest of the skies' did so, if in a different manner: the spirit of adventure persisted, as well as exhilaration in risk-taking and daring, but not within the calm of timelessness symbolized by the mountain. Here, in the sky, the spectacular, the theatrical, counted; it was part of the appeal of the flying ace. This was the kind of drama mountaineering could not provide. Once more, the theatrical was easily annexed by the nation as part of its self-representation, exploited for such purposes after the war. For example, Italo Balbo's flight across the Atlantic in 1933 with 24 planes was meant to alert the world to the new spirit that governed fascist Italy. But in France as well, the 576 aeroplanes which flew overhead on Bastille Day in 1935 used the skies to intensify the theatrical sensibilities of state politics.

Aviation was used to enhance state power after the 1914–18 war, for it still fulfilled some of the functions it had provided for the 'myth of the war' experience. In the defeated nations it encouraged the enthusiasms that had moved the youth of 1914, helped create a new soldierly man to counteract the supposed degeneracy of the German republic, and promoted the 'new man' image Italian fascism hoped to create. Since Germany was not allowed to have an air force until a decade after the peace, gliding took its place and became a popular sport, with overtones of nationalism and the heroism of aviators in war. Eventually Hermann Goering himself drew attention to the continuity between the wartime aces and modern times; indeed he owed his rapid rise in the Nazi movement to the fact that he had been the last commander of the von Richthofen squad. Sacrifice for the nation and the virtues of manliness were the two main themes from the war experience now being taken up by the 'conquest of the skies', whereas individualism and self-realization, which had also been valued earlier, were considered unimportant.

The myth of wartime aviation was fully exploited by the National Socialist and Fascist dictators. Mussolini, for example, used his own skill as a pilot to enhance his image, while Hitler was pictured as having saved the day through his courage during a stormy flight. A National Socialist book, which attempted to inform young men whether or not they had an aptitude for flying, told them that their souls would have to be at one with the aeroplane, and that comradeship and decency were the basic virtues needed, together with iron discipline and the ability to give one's all for a great cause. The myth of the conquest of the skies was co-opted by National Socialism, not only through the person of Hermann Goering, but also as exemplary for those virtues the dictatorship needed for its own ends, as creating the kind of citizen it desired and in preparation for the war to come.

By contrast, the themes of individualism and self-realization were what attracted people in other nations to the 'conquest of the skies'. Man alone in the infinite spaces had not proved to be a military option in the long run, and late in the Great War planes always flew in formation, but it was nevertheless an ideal that retained the power to move millions. The adulation Charles Lindbergh received everywhere he went after his lone flight across the Atlantic is proof of this. It is quite possible that this adulation had something to do with the feeling that this flight too camouflaged the war, or rather its consequences: the cripples on the streets, the bereavement, the empty sleeves — here these unsightly reminders were overcome. But Lindbergh as a hero above all incorporated all those qualities the air ace had symbolized during the war, especially individualism in an ever more depersonalized world, and manliness, which so many of the new generation had missed in the war.

A continuity existed, at least in Germany, between the generation of 1914 and those who came later and never experienced the reality of war. The thrill of aviation mediated between these generations, and Lindbergh not only personified those virtues many a youth desired, but through his comportment and even his looks provided an example of the man of steel: he was supposedly modest, quiet and virile, yet had himself under control. He was the ideal hero for the times, exemplifying by his deeds, appearance and character the very opposite of a world grown old and tired.

As important as Lindbergh, Antoine de Saint-Exupéry transmitted the myth of flying into the 1930s, when his *Wind, Sand and Stars* and *Night Flight* became world best sellers. Here are the same themes once again, with the ideal of 'overcoming' one's fears writ large, rising above oneself; and, although human life was said to be invaluable, something does exist which has a higher value. Death too is overcome, as we read in *Night Flight*; the human body is exchanged for something eternal. Here also hardness easily becomes brutality. Saint-Exupéry was an allied flyer in the Second World War and was shot down in 1944. He was not a fascist, nor even a particularly pronounced nationalist. Nevertheless, he emphasizes pre-industrial virtues and an unabashed elitism. Certain types of warfare, especially in the air, were said to be ennobling compared to the ant-like existence imposed on the inhabitants of an industrialized continent.

The 'conquest of the skies' was a myth, which began with the Great War and captured as well as transmitted the spirit of 1914, lost after the war began. This myth fulfilled the need to mask the reality of war, projected so-called old fashioned virtues associated with manliness and informed society's ideals. Popularized by writers, statesmen and the media, the myth penetrated all levels of culture and popular consciousness. The actual behaviour of aviators played little part, no more than the reality of warfare, except that in some of the pilots, like Boelcke or Lindbergh, reality and legend coincided to a singular degree, making them all the more effective as exemplars of what it meant to conquer the skies.

How effective was this myth in the long run? Obviously it was alive in the interwar years as we have seen, furthered by nations expecting or preparing another conflict. And yet, in 1939, all such myths, all attempts to mask war, failed. To take the most extreme example, when

Hitler drove to the Reichstag to declare war, he received no applause at all and this for the first time during his dictatorship. He himself realized the failure of the myth of the war experience, for he thought it necessary, in order to start a war, to stage a fake assault to make it look as if Germany had been attacked by Poland. What had worked in 1914 no longer worked very well in 1939, in spite of the tempting dream of conquering the skies. Perhaps in Nazi Germany the aeroplane itself was one cause for this failure: the constant air-raid exercises were counterproductive, making the population aware of what might await them.

Not all was lost, of course, but after the Second World War the heroic no longer aroused much enthusiasm. As far as Germany was concerned, some would not let go, such as Heinz Konzalick, who in his best seller *The Doctor in Stalingrad* (1958) focused on the SS as true men who had fought a rearguard action in the east. Perhaps more to the point, the very popular pamphlets, the *Landserhefte*, for a while made war out to be a great but deadly adventure, the proof of manhood once again. But today the tone of the *Landserhefte* has changed, if not completely, for in a recent issue the air ace is still a lone warrior, even though surrounded by other war planes, and some of the thrill is still present. Now youth is said to yearn for life, not death, however glorious, and the hero is said to fly as one destined for death, through a sky that has become a living hell. But such publications are now fringe phenomena without state support. The mountain film took longer to lose its theme of mission and conquest, but by the 1970s this too was past.

Wars went on, becoming perhaps even more numerous after the Second World War, but in Europe and the United States the 'myth of the war' experience fared less well. To be sure, the search for heroes continued (there was a revival of interest in Lawrence of Arabia) — heroes in wide open spaces and thrown upon their own devices. The search for the old-fashioned virtues the aviator had symbolized also continued, but was no longer bolstered by war. Whether the quest for the test of one's manliness and all that this meant goes on seems dubious as well. In any case few today would regard war as such a test; the Second World War was too destructive and affected not just the military but the whole population.

But all such continuities or discontinuities have not influenced the waging of war. Does the reality of war no longer need masking? As recent wars are mostly civil or colonial, are they so different that they can dispense with such devices? As we look at the Balkans or Russia today, is nationalism by itself enough to make the death and deprivation of war bearable? The volunteers of 1914 were also nationalists, but as the conflict unfolded more concrete signposts were needed to overcome the reality of war; and war myths, like the cult of the fallen soldiers or the 'conquest of the skies', played a crucial part in keeping the memory of the glory alive despite all that destruction.

The myth of the war experience has not yet found its historian. Yet the historical function of the 'conquest of the skies' may well not have ended after all: the stereotype of the ace, the virtues he projected, the attraction of the wide open spaces against the littleness of life, are still with us, even if they have taken on many other, more up-to-date forms. We should not cease to examine the political rhetoric and the popular culture of our time for traces of such myths, for they have too often served to disguise the reality and to cleanse the memory of war.

10. The Role of the Mass Media in Modern Wars

Jo Groebel

Modern wars need the mass media; the mass media need wars. These statements have to be explained: immediately after the ending of the Gulf War an Israeli general described television as one of the most important means of military strategy. Most governments would have difficulty engaging in a war without the support of the population. This support depends on the belief in the necessity of military action and its emotional rewards. As the majority of the younger generation have had no personal experience of war, their knowledge stems primarily from mass communication: modern media play a substantial role in the construction of reality as well as in attitude formation in the audience. This is especially true of nations with a high level of electronic communication systems, such as the USA. It has now been demonstrated and is common knowledge that the outcomes of American presidential elections are determined not only by political programmes but at least to the same extent by a candidate's performance on the screen, i.e. his or her audiovisual image (Jackson-Beeck and Meadow 1979, and many others). This concept of 'image' and its importance for public opinion can also be applied to wars. While propaganda has always played a crucial role in military activities, it has reached new professional levels with the occurrence of global television and the development of increasingly differentiated public relations strategies.

While the American public already had a generally pro-action attitude before the Gulf War actually began, the Kuwaiti exile government wanted to reinforce this trend and extinguish the last doubts. In this situation the world's biggest PR firm was hired to develop a systematic communication strategy. In the first stage, it tested what kinds of images and symbols would be most likely to elicit emotional, pro-war reactions from the audience. Not surprisingly, babies and young girls turned out to be especially 'emotional' topics. As a consequence, a video tape was quickly produced and distributed wherein a young girl reported in tears how she had witnessed Iraqi soldiers occupying a Kuwaiti hospital and removing premature babies from incubators they intended to take back to Baghdad.

The video was shown on all the major news programmes and led to further public and political support for the Bush administration's plan to start the war. After the conflict ended, it turned out that the young 'witness' had been the Kuwaiti ambassador's daughter who had not been outside Washington for several years. At the same time, a medical commission visiting Kuwait was surprised to find that the hospitals did not have any shortage of incubators. So, despite other cruelties by the Iraqi army, *this* particular story had obviously been made up.

While this event may not be an adequate criterion for evaluating whether the war was 'justifiable' in any respect, it demonstrates the importance of communication strategies in modern wars.

This role of communication as a strategic tool meets a media demand for any spectacular event — with the occurrence of mobile broadcasting, live television has become a 'must' in the modern media world. In this context, the Gulf War was probably the first war in history to be announced in television guides. It had its own logos and its own jingles, and the audience could find its beginning announced in the programme press. The war actually started during (US) prime viewing time.

During the war itself reporting was heavily censored and the military actions were presented as video games without any further casualties. Showing victims might have created stronger opposition to the military intervention. American news reporters partly abandoned their original role as 'neutral' observers and used subjective language like '*We* attacked *them*', thereby reproducing the enemy image-terminology applied by military PR agencies.

The reports concentrated on the actual attacks and gave very little background information on the historical development of the conflict. Thus, most Americans who were asked in a survey knew many details about 'Patriot' missile technology but were hardly informed at all about the geographical and cultural situation of Kuwait and Iraq. It was a war of 'good' against 'evil'.

With a different kind of dramaturgy the arrival of American soldiers in Somalia also was a major example of 'modern' news reporting. Again, the timing of the arrival was centred around the programming of the American broadcasting media. When the army arrived on Somalia's beaches in full battledress it was facing not the 'enemy' but hundreds of journalists, cameras, professional lighting, a scene not unlike the staging of a Hollywood movie. Faced with this situation the soldiers obviously felt some pressure to behave in a 'camera-fit' manner. To act as heroes they sought out potential 'enemies', bound them up and, with advice from reporters on the right gestures and positions, posed in front of the cameras. It seemed as if they were trying to emulate the performances of actors in a film. Nevertheless, the intervention, which was started to reduce famine, was welcomed by the American public.

But Aidid (the Somalia 'Liberation Army' leader) also knew how to handle the media. When the dead naked body of an American soldier was cruelly beaten and taken through the streets, the scene was recorded on video and distributed to the international media. With the need for spectacular audio-visual events, many news reports featured the even crueller details of the scene and thereby promoted Aidid's wish for global attention. The shocking images destroyed any illusions people may have harboured about the death of a 'glorious' hero. As a consequence the American public changed its opinion and became far more critical of the US engagement in Somalia. Thus, even a single image can have the power to influence attitudes towards military activities.

The Classical Concept of the 'Enemy Image'

Communication has always played a major part in army strategy. During the First World War French and German soldiers voluntarily marched to their certain death and were applauded by enthusiastic national press propaganda. One of the earliest systematic studies of the influence of the mass media describes these processes (Lasswell 1927). During the German Nazi period propaganda strategies were increasingly professionalized and findings of academic psychology were applied (Groebel 1992).

In that context, one major element has always been the use of enemy images. While one's own side is described in terms of moral motives and heroic behaviour, evil and cowardice are attributed to the enemy. By enlarging the differences between 'friends', and 'enemy', any inhibition about attacking the other side is decreased.

In more general terms, creating an 'enemy image' can be described as follows: the description distinguishes between intention, message, target, means, process, effect: to reach acceptance of or trigger an attack against another (intention), a stereotyped, simplified and negative image (message) of the characteristics, motives, intentions and behaviour of the other person or group (target) is created in which emotional or pseudo-informational mass media elements (means) apply selection, exaggeration and fake (process), with the consequence that the social distance between the two sides is increased (effect).

One may also wish to include 'sender' and 'receiver' variables and their interaction in this description, like military leaders or propaganda experts on the one side and a population/global audience or average soldier on the other. The enemy image as described above is exclusive to humans. Non-human primates may also discriminate within and between groups and may even manipulate others in order to create negative attitudes towards a specific target. But it seems that only humans are able to create enemy images through symbolic representations without any face-to-face contact, thus permitting mass propaganda directed at distant goals. Of course, these processes are enhanced by the application of fundamental psychological patterns like fear of strangers or in-group/out-group behaviour.

All in all, the creation of an enemy image — simple as this image itself may be — is a highly complex process involving different levels of social and psychological reality. In this, it resembles the complexities of aggression itself (see, for example, Attili and Hinde 1986; Groebel 1983; Hinde 1993; Groebel and Hinde 1989b). Actually, the mere creation of an enemy image can in itself be regarded as a form of (psychological) aggression.

From the description given above, the process of creating an enemy image has diverse features :

At least, three different *individuals/groups* are involved: the communicator, the audience and the target. Each has specific characteristics which interact with those of the others. Social psychologists have described many of these characteristics, such as credibility of the communicator and intelligence of the audience.

In this context, a major *precondition* for creating an enemy image seems to be that the audience should already have prejudices against the target group, as is often the case with people from different ethnic and cultural backgrounds. Typically, these differences are attributed to 'bad intentions', 'stupidity' or 'fanatical mentality', in, for example, the evaluation of the non-verbal behaviour of Arabs by Westerners. With a lack of personal communication, this tendency can be reinforced as no correction of the original impression takes place.

Preconditions include:

- the already mentioned prejudices;
- lack of personal communication;
- the general tendency of individuals to base evaluations of others on a few simplified impressions;
- an often dichotomous image of the world (good guys/bad guys);
- fear of strangers; and
- a fascination for 'evil', in which even the 'enemy' on the screen becomes a favourite: people 'liked' to see Saddam Hussein on TV and to get upset by the image. This can be explained, without necessarily resorting to the social psychological approach of in-group/out-group thinking, in terms of arousal needs where no real threat is present.

Apart from the already mentioned *purpose* to trigger a war, the establishment of an enemy image can also serve

- to justify any form of violence;
- to draw attention away from one's own side's faults and problems;
- to strengthen the in-group by defining an out-group; and
- to attribute the causes of frustration to a clearly defined outside source.

The following *mechanisms* apply:

- The other group is described as a homogeneous mass; individual differences are neglected.
- The others' complexities are reduced to simple, 'primitive' features.
- The target group is labelled in diminutive or ridiculing terms.
- In extreme cases, the other group is 'dehumanized', given animal or abstract object names and connotations in order to facilitate the willingness to attack them (typical characterizations include 'pigs', 'rats' and 'evil empire'). Visually, existing characteristics are exaggerated in caricatures and propaganda films and posters.
- A whole group of people is personalized into one simple picture of *the* enemy: Russians become *the* Russian, Americans *the* American. This applies even more if a single leader is at the centre of attention and, for example, Saddam becomes the epitome of all Arabs, or, vice versa, Bush the epitome of all Americans.
- Aggressive intentions are attributed to the other side; the in-group only defends itself.

- One often can find a mirror image: both sides use the same images and stereotypes, a favourite being Hitler. Bush identified Saddam with Hitler, Saddam did the same with Bush.
- Using formal features of the mass media to create an emotional climate (dramatic music, rousing voices, special effects).
- Facts are selected and put in a simplified context to create a one-sided cognitive belief system. To prove the necessity of 'preventing' an attack by the other side, the number of its weapons is exaggerated or strategic plans of the enemy are 'discovered'.
- Casualties on the other side are not reported or are put in abstract terms, so that any possible empathy with victims is reduced.
- Subtle forms of enemy communication include the verbal or visual construction of indirect causal links between the other's ethnic characteristics and specific behaviour patterns ('Chinese kills woman').

Many of these mechanisms are often used by the modern Western media. However, in crisis situations similar elements return even to the 'sophisticated' press, as in many reports during the Falkland Islands conflict or the Gulf War. In dictatorships or actual battle areas they are still common, as in the war in Bosnia.

Psychological Processes in Modern War Communication

A major shift in 'war communication' occurred with the omnipresence of violence in the audiovisual media in general. Physical aggression is an important ingredient of television and video. On average, American programmes present ten acts of violence per hour per channel 24 hours a day. European television scores lower but still contains approximately five acts of physical aggression per channel per hour (Groebel and Gleich 1993). Though the majority of these scenes stem from fictitious programmes, news 'shows' have also become part of the violent diet. A special programme category has been created that merges real content with the formal features of fictitious programmes, so-called 'reality TV'. Without further analysis of the background one can watch the most spectacular scenes from real criminal or military events presented in a highly dramatized form, including background music, video-clip aesthetics and sensationalizing 'off' voices. In addition, some real events seem to be tailored around the patterns of fictitious stories. The Tonya Harding–Nancy Kerrigan 'ice queen' conflict before and during the 1994 Winter Olympics had all the elements of a Hollywood movie plot and was dealt with by the media in exactly that way. The high ratings of 'reality TV' programmes demonstrate their attraction. In fact, war seems to be so especially 'media fit' that an exclusive 'military' channel is in the making for American television, which will contain 24 hours of war reporting from all over the world. Even without such a channel, switching through the programmes demonstrates the omnipresence of war on the screen.

Can this be regarded as led solely by a normal desire for information? The reality is more complex. Violence and action of any kind are particularly attractive for programme makers and the (male) audience alike as they arouse special attention by their visual and auditory intensity. At the same time, they draw a picture of a world in which complex problems can be solved by simple means. Furthermore, the idea of the strong hero (see other chapters in this book) may serve the identification needs of many adolescent and adult males. This is especially true when war reporting concentrates, not on the casualties and fatal consequences of military action, but on our fascination with the precise computer army technology that has partly replaced the old-fashioned hero concept. But even images that present the cruel side of war do not automatically create a negative attitude towards violence. Apart from their voyeuristic attractions, showing the same or similar war images over and over again may create habituation and decrease sensibility towards these images. The audience gets 'used' to the Bosnian war; scenes from the Vietnamese War are now sold as postcards. The quality and quantity of war communication has changed: it is a normal part of our lives.

Thus war has become a major ingredient of an overall visual programming system. What processes are involved and what effects do they have? First, one has to distinguish between intended war communication, as described in the section on 'enemy images', and communication in which war is used as part of the programme maker's entertainment policy. In the latter case, the reporting serves either the investigative goals of the journalists, or the market orientation goal of reaching as large an audience as possible. Even then, the *effects* (social distance, dehumanization) can be the same as for the intended enemy image communication. This is especially true of war as part of entertainment. While there are outstanding examples of anti-war films, such as *All Quiet on the Western Front*, *Die Brücke* (The Bridge) and *Apocalypse Now*, most fictitious war films presented on the screen deal with the good guy/bad guy stereotype and glorify the individual hero or a 'fascinating' technology, as in films like *Rambo* or *Red Dawn*.

These films may only attract a specific audience, i.e. the male adolescent, but for this group they provide an 'implicit war education' as opposed to one centred around peace. The viewers may not immediately feel attracted to the subject of military engagement, but 'learn' that violence is an adequate and fascinating means of gaining control, solving conflicts and guaranteeing an upright 'moral' position. Moreover, soldiers in real wars identify with the models they have observed in Hollywood movies. News reports from Bosnia show Serbian soldiers posing with their weapons and declaring themselves to be the new 'Rambos'. Military action is perceived as an adventure in which real men prove their 'guts'. The messages inherent in most fictitious war films promise rewards on six different levels:

- *social*: one can become a hero;
- *economic*: one can gain material or psychological profit;
- *emotional*: war means excitement, adventure;
- *cognitive*: war is necessary to gain control in a chaotic world;
- *physical*: the physiological and motor activities are perceived as pleasant and exciting; and
- *moral*: acting/dying as a good guy means moral payoff.

Thus, the myth of war is continued even in films that are not explicitly propaganda-oriented.

With the overall importance of violence in the media, war reporting and military fiction become part of a system in which there is a high risk of aggression-increasing effects on the audience (see the 'state of the art' of research on television and aggression, for example Huesmann and Eron 1986, Comstock 1991, Groebel and Gleich 1993).

Applied to the military context, the following model describes the different effects of violent television. Most of the elements in the model have been empirically tested (Groebel and Gleich 1993) though these empirical tests can not be reported here.

Moderated by audience variables, especially *gender* (primarily males are attracted); *motives* (sensation-seeking; mood-management orientation needs), family and upbringing (see Hinde 1993); and *culture* (see Hinde and Groebel 1991), the effects of media violence on the audience are:

	Arousal	Belief in evil world	Habituation
presentation of war in the media	fictionalization of war reports	acceptance of images of enemy	media construction: war normal
has effect on perception that	war is exciting and rewarding	war is necessary	war is unavoidable

There may even be an immediate 'imitation' of behaviour related to military action among children, for example with toy guns. And, as already stated, soldiers may feel attracted to fictitious war heroes and their activities. However, the major effects of war representation in the media lie in the acceptance of the 'hedonistic' qualities of war and the belief systems centred around the necessity for armed conflicts. In the long run, the dramaturgical demands of the media and their effects on the audience interact: war, due to its high arousal qualities and immediate visual 'action', is more 'media fit' than peace. It receives more attention and is thus over-represented as compared with other means of conflict resolution, which are more complex. In addition, the media, realistically, have to report about failed peace negotiations more often than about successful ones, as in the Bosnian war. The resulting message: 'War is normal'.

With audience ratings playing the dominant role in television programming, war reporting has also become part of an industry. Visual effects and being 'live' on the screen have become more important than background analysis. There are of course differences between television cultures and individual channels, but with the increasing influence of global visual agencies the available news material has become more uniform all over the world. One can even observe a partial 'fictionalization' of war reports in which it is difficult to distinguish between the dramaturgy of a movie and a news report. During the Gulf War the use of special-effects trailers, sound, jingles and videogame images made it difficult to imagine that real casualties were taking place during the event. In addition, one could buy war simulation games that enabled the audience to 'play' the attacks at home. The fact that a fictional reworking for the screen of the events of a conflict is already on the way adds further to the construction of a world in which reality and fiction merge. Cognitively, the audience may still be able to distinguish between the two, but the emotional reactions will be subject to confusion and habituation (Groebel and Gleich 1993). These are the 'new qualities' of the modern media world as compared with the role of communication in the pre-mass-media world. Although one should not forget the constructive possibilities of the media — information and the growth of cultures — the (unintended) effects of news and fiction on the promotion of war have to be acknowledged.

11. Patriotism And Nationalism: Two Components of National Identity with Different Implications for War and Peace

Seymour Feshbach

The history of our species can be conveyed in terms of the warfare that has occurred between tribes, nations, empires, religious groups and other social and political entities. It would seem that a fair reading of history and of the contemporary behaviour of nations serves only to reinforce Freud's conception of humans as being driven by biologically rooted aggressive and sexual impulses that need to be controlled and regulated by the rules and sanctions of civilized society. Nevertheless, an instinctual, biological model of human aggression, however compelling, is not an established fact. It remains as one view of the basic source of human aggression and violence. While one can cite a litany of historical events documenting human violence and inhumanity, one can also cite historical events that support a view of humans as basically cooperative creatures who prefer amicable relationships with other humans.

This chapter is guided by a model of human behaviour and a view of human aggression that assert that humans are capable of a wide range of behaviours — from love and cooperation to anger and destruction, and that while nature may contribute to dispositions to act in particular ways in particular situations, nature does not 'dictate' human behaviour. This view of human nature may provide a more optimistic perspective of human potentiality and the future than do the Freudian or biological models. However, it is not intended to minimize the enormous difficulties entailed in establishing political structures and interpersonal and inter-ethnic attitudes and relationships that will significantly reduce the willingness to resort to war as a means of achieving political objectives. It does suggest the possibility of fostering societal arrangements that will enhance the human potential for cooperation, compromise and other non-warlike responses to conflicts between nations, and between ethnic and religious groups.

Social and psychological analyses indicate that the determinants of war are varied and complex. In this chapter I focus on individual differences in support for military policies, including the readiness to go to war. Population and economic pressures, ethnic and religious differences, propaganda and group conformity are, of course, of major importance. Nevertheless, I believe, and hope to demonstrate, that it is also useful to examine individual differences in the degree of support for war and war-related options. Although there are powerful cultural forces that drive nations, ethnic and religious groups into military conflicts, these pressures are usually accompanied by individual differences in degree of support for military options. The examination of individual differences provides one avenue for discerning the psychological factors that influence support for 'hawkish' versus 'dovish' national policies.

Individual Aggression and Attitudes Towards War

I began my exploration of individual differences in the factors that lead some people to favour hawkish and others dovish military policies by examining the role of individual differences in aggressive tendencies. Some people manifest considerable aggression in their interpersonal interactions while others manifest very little. There are some individuals who are easily angered, who are ready to attack when frustrated and others who are more even-tempered and who seek more amicable methods of resolving conflicts. I am not concerned here with the degree to which these individual differences in aggressive tendencies are biologically based. The principal question at this point is the link between individual differences in interpersonal aggression and the degree of support for war. Is the degree to which an individual is ready to support hawkish, aggressive national policies primarily a reflection of the degree to which which he or she can be described as having an 'aggressive' personality?

In our earliest studies addressing this question (Feshbach, Kandel and Haist 1985), we found, somewhat to our surprise, only marginal relationships between measures of individual aggressiveness and the degree of pro-war and pro-nuclear armament sentiments. In further studies, a series of vignettes describing provoking behaviour by foreign powers was presented to a sample of college students, who indicated their degree of support for the United States going to war in each of these situations. A significant but only modest relationship was found between the readiness to go to war and the degree of individual aggressiveness. It appeared on the basis of this series of studies that it would be appropriate to look elsewhere for a major source of individual differences in the readiness to support militant, hawkish national policies.

Patriotism and Nationalism

We then turned to an examination of the role of national orientation in determining the degree of support for war and related policies. One's relationship towards one's nation entails a complex set of attitudes, feelings and values and is not readily encompassed by a single concept such as 'patriotism'. In addition, the scientific analysis of this complex relationship has been further complicated by changing social attitudes towards its moral status. Following the Second World War, the term 'patriotism' had acquired a negative connotation for academics and other members of the intellectual community in the United States and, undoubtedly, in other nations as well. Patriotism became associated with ethnocentrism and aggressive military pursuits and was contrasted with the virtues of 'internationalism'. Yet patriotic-nationalistic feelings can be among the noblest as well as among the most destructive of human impulses. These feelings are associated with positive virtues such as heroism, loyalty and self-sacrifice as well as such negative attributes as hostility towards out-groups, self-aggrandizement and unrestrained violence. In order to examine the relationship of patriotic-nationalistic sentiments to support for militant, aggressive national policies, it is important first to separate, in so far as possible, these positive and negative components.

To this end, we developed a comprehensive questionnaire measure of one's relationship towards one's nation (Kosterman and Feshbach 1989). Analysis of this instrument indicated that it consisted of a number of components or factors that could be separately measured. For our purposes, we shall focus on the two most important components. The first of these consisted of such questionnaire items as 'I love my country'; 'I am proud to be an American'; 'In a sense, I am emotionally attached to my country and emotionally affected by its actions. ' We labelled this component 'Patriotism'. The second component consisted of items such as 'In view of America's moral and material superiority, it is only right that we should have the biggest say in deciding United Nations' policy'; 'Other countries should try to make their governments as much like ours as possible.' We labelled this second component 'Nationalism'. The assignment of verbal labels to factors typically entails conceptual difficulties, even arbitrariness, in that the labels may have common language meanings in addition to those defined by the factors. Political scientists and historians, in particular, may find that the ascription of these labels fails to do justice to the complexity and nuances of patriotism and nationalism as manifested in different historical periods and different political structures. Such a terminological debate could be eschewed by simply using the labels 'Factor 1' and 'Factor 2'. However, there is a linguistic awkwardness in such usage, and psychologists typically assign conceptual labels to the dimensions derived from factor analyses. The essential point is a recognition of these different components of national identification and sentiment, such recognition being the first step in the determination of possible differences in their properties.

It is an empirical question as to whether analyses of national feelings carried out in other countries would yield similar components of patriotism and nationalism. However, it is noteworthy that an analysis of a Japanese version of the questionnaire instrument administered to a very large number of college students in Kyoto yielded components of patriotism and nationalism almost identical to those found in the US sample.

Patriotism and nationalism are related to each other. However, the relationship is not a perfect one. Patriotism pertains to the identification with and attachment to one's nation, while nationalism refers to feelings of national superiority and the desire for national power and control. There are many individuals who are high in patriotism and low in nationalism, and even some individuals who are very nationalistic but obtain low patriotism scores. Moreover, and most importantly, patriotism and nationalism have different functions and consequences. The thrust of our research programme has been the demonstration of differences in the antecedents and consequences of patriotism and nationalism, despite the positive correlation that obtains between these two dimensions of national identification. And of particular importance is the demonstration of differences in the implications of these dimensions for war and peace.

The distinguishing feature of patriotism is the element of positive attachment. The attachment entailed in a patriotic relationship has a functional similarity to the attachment of children to their parents in a family relationship. We have subjected this hypothesized relationship to empirical examination. As a first approximation to the assessment of childhood attachments, we developed a self-report questionnaire measure of early attachment to one's mother and early attachment to one's father. This measure can be influenced by memory distortions. At the same time, however, we have shown that it is highly reliable and that in completing the questionnaires, individuals distinguish between their current feelings of attachment to their parents and their early feelings (based on their earliest memories). The measures of early parental attachments and of patriotism and nationalism have been administered to large samples of college students in several different studies (Feshbach 1987; 1990; 1991). A positive relationship has consistently been obtained between patriotism and reported early attachment to the father; that is, for both men and women, those individuals reporting stronger early attachment to the father also obtained higher patriotism scores. In contrast, the measure of nationalism was unrelated to paternal attachment. It is of interest that a similar pattern of findings was found in a sample of Japanese students, except that for the Japanese patriotism was found to be related to maternal attachment as well as paternal attachment.

Three aspects of the findings yielded in these studies warrant emphasis. The first is the relevance of the phenomenon of attachment to patriotism. There is a substantial amount of psychological literature attesting to the positive developmental consequences of a loving, secure early parental attachment. Perhaps patriotism, unconfounded by nationalism, may have similar positive features. The second aspect of the research findings that is noteworthy are the differences obtained in the relationships found for patriotism and for nationalism. We have carried out a series of investigations that demonstrate differences in the properties of patriotism and nationalism, several of these studies employing experimental as well as correlational methods. Thus, one experiment entailed having individuals either engage in thoughts of permanent separation from a parent or in neutral fantasies, following which they completed measures of patriotism and nationalism. Women who had been instructed to fantasize separation from the father manifested a significant increment in their patriotism scores. Interestingly, and further demonstrating a functional difference between patriotism and nationalism, men who were asked to fantasize separation from the mother had significantly higher nationalism scores than the controls (Feshbach

1992). In a quite different experiment demonstrating patriotism–nationalism differences, it was found that listening to a US patriotic anthem resulted in a significant increment in a sample of US students on the patriotism but not on the nationalism scale, while listening to martial music affected the latter but not the former (Feshbach forthcoming).

The third aspect of our research programme, and most germane to the current issue, concerns individual difference factors influencing support for hawkish versus dovish national policies. Strong relationships have consistently been obtained between nationalism and measures of support for nuclear armament and for military actions. While patriotism is also related to indices of militarism, the relationships are much weaker than those obtained for nationalism. In this regard, an ironic outcome of one study was the finding in a United States sample that nationalists were more ready than patriots for the United States to go to war in various provocative situations, but the patriots were more ready than the nationalists to risk their lives in the event of war in these same situations.

National Feelings, Sociobiology and War

Our focus thus far has been on delineating the components of one's relationship to one's nation, and determining the extent to which these components are germane to individual differences in support for military solutions to conflicts between nations. We have distinguished two components of this relationship — patriotism and nationalism. These components do not adequately describe one's relationship towards one's nation, which entails a complex constellation of other feelings, aspirations, perceptions and values. However, the distinction made between patriotism and nationalism is useful as a first approximation to the analysis of national identification, feelings and commitment. An important consequence of this distinction is the finding indicating that it is the nationalism component of the relationship to one's nation that is most strongly related to support for militaristic, war-oriented policies. This is not to say that the patriotic component is unrelated to militarism. However, the relationship is a modest one, particularly when one separates patriotic and nationalistic feelings.

We have had little to say about the antecedents of nationalism, as it has been defined here, focusing instead on the importance of early paternal attachment for the development of patriotic feelings. The attention given to early attachment suggests that patriotism is related to early social influences. While this analysis does not exclude the possible role of biological factors in either patriotism or nationalism, biological factors are certainly given little emphasis in our approach to the understanding of national feelings and their origin. There are competing views that see national feelings, suspicion of other groups and warfare as a biologically ingrained behavioural pattern. Shaw and Wong (1989), in their discussion of nationalism, contend that 'warfare propensities are deeply entrenched in human nature' (1989: 17). They maintain, on the basis of genetic and evolutionary principles of inclusive fitness and kinship selection, that individuals have inherited strong tendencies to defend and expand the power of their kinship groups and to be fearful and suspicious of 'strangers' (members of non-kin groups). These tendencies are manifested in the range of sentiments and behaviours that are connoted by the terms 'nationalism' and 'patriotism'. Shaw and Wong recognize that nations may consist of diverse groups who are as genetically dissimilar from each other as they are from members of other nations; and they invoke psychological mechanisms, such as identification and perceived similarity, to account for patriotic attachment to groups that have little kinship similarity to the individual. The consequence of this extended identification is that the biological proclivities to expand the interests of one's kinship group through war or other expressions of power become implicated in one's relationship toward larger and genetically dissimilar groups.

Shaw and Wong's evolutionary model of humanity's propensity for warfare, while congruent with much of what has been historically recorded about warfare, also has significant difficulties and ambiguities. Allusion has already been made to the necessity for proposing a mechanism to account for nationalism in nations made of genetically diverse groups. In addition, the violent propensities manifested in civil wars and in religious wars, in which aggression is directed toward genetically similar individuals, cannot easily be reconciled with a genetic model that is based on inclusive fitness and kinship selection. Why perceived similarity should supercede biological similarity or why the perceived similarity of religion should supercede that of nationality is not evident. Shaw and Wong are not alone in not being able to offer an adequate explanation of internecine wars. There is little in the way of theory and research that helps us to account for the intensity and fervour of violence where ideologies rather than territories or kin are at stake.

Patriotism, Nationalism and Values

Our own analysis of the relationship of the individual to his or her nation has placed great emphasis on the distinction between patriotism and nationalism, and on the differences in the functions and consequences of each of these components of national identification. We have suggested that many of the negative consequences of national identifications are borne by the nationalism component. And we have proposed that the essential feature of the patriotism component is the element of attachment.

Although we have had little to say about the antecedents of nationalism, our view is that psychological and cultural variables are critical to an understanding of both patriotism and nationalism. What is it that shapes human characteristics so that some individuals are highly patriotic, others highly nationalistic, some both and some neither? We do not yet have an answer to this ambitious question. The work on early attachment is germane. Parental attitudes, schooling and other cultural factors undoubtedly play an important role in the development of patriotic and nationalistic sentiments. Prior studies that we have carried out suggest the influence of personality factors. Recently, we have completed a study of variables more proximal to patriotic and nationalistic sentiments — namely, value differences. One can view one's relationship towards one's nation in terms of the satisfactions provided by the nation — more broadly, the values and goals of the individual that the nation helps realize. These values may be economic, political, religious, military or cultural. The degrees of patriotic attachment and of nationalistic sentiment are hypothesized to be in part a function of the respective values and goals held important by individuals high in patriotism and those high in nationalism, and of the extent to which these values (and goals) are perceived as being realized by one's nation. Of particular interest are differences in values associated with patriotism versus nationalism, and the import of these values for peaceful or militant attitudes towards other nations.

To explore these issues, a standard measure of values, the Rokeach (1973), was administered to several hundred college students, along with the measures of patriotism and nationalism. In as much as the values assessed by this measure are very broad in nature, a value (and goal) inventory more specifically linked to expectations of one's nation was developed and also administered to the sample. The latter inventory included the following values, which the respondents were asked to rank in order of importance: democracy; ecologically sound environment; economic security; economic opportunities (for wealth); freedom of speech and press; freedom of religion; military defence; military power and dominance; protection from crime.

These values are not exhaustive. They do not include the more subtle gratifications associated with national attachment such as a shared history and cultural experience. They are not intended to embrace all of the meaning and import of one's identification with and relationship to one's nation. They are assumed to be independent of the process linking early paternal attachment to patriotic feelings.

To simplify interpretation of the data, foreign nationals were eliminated from this analysis. The correlations of the value rankings with patriotism and nationalism indicated that while individuals high in patriotism and in nationalism assign higher values to military defence, only the nationalism scores are significantly associated, as one would expect, with higher value rankings ascribed to military power and dominance. Interestingly, while both high patriotism and high nationalism are associated with the ascription of low rankings to freedom of the press and of religion, high patriotism is associated with higher values for political democracy, and high nationalism with higher values for economic opportunity.

The measures of the extent to which these values were perceived as being realized in the USA were, in general, positively associated with patriotism and nationalism. Some interesting differences also emerged here. Freedom of speech and press and freedom of religion realization scores were more strongly associated with patriotism than with nationalism (correlations of .42 vs. .25, and .26 vs. .14, respectively); perception of degree of protection from crime was significantly associated with nationalism only (.20); and the military power realization scores were more strongly associated with nationalism than with patriotism (–.31 vs. –.19). The signs for the latter correlations are negative, indicating that nationalism, and patriotism to a lesser extent, is associated with a tendency to perceive US military power as less than it should be.

To sharpen possible differences in values between patriotism and nationalism, the sample was divided at the median on the patriotism scale and also on the nationalism scale, thereby constituting four groups: low patriotism-low nationalism, low patriotism-high nationalism; high patriotism-low nationalism; high patriotism-high nationalism. The most striking finding that emerged from this analysis was the difference in the values assigned to political democracy. Respondents who were high in patriotism but low in nationalism placed the highest value on political democracy, significantly higher than that placed by the low patriotism-low nationalism group or the low patriotism-high nationalism group. Moreover, the value score of the low patriotism-high nationalism group was significantly less than that of the high patriotism-high nationalism group. In addition, the lower evaluations of freedom of speech and press and freedom of religion reflected in the correlations with patriotism were no longer significant when nationalism was separated from patriotism.

The analysis of the Rokeach terminal value rankings also yielded differences of interest, although in this instance the values were more closely linked to personality attributes. We found that individuals high in nationalism tended to value more highly 'a comfortable life', 'pleasure' and 'social recognition' whereas the correlations of these values with patriotism were negligible and non-significant. The nationalists also valued 'national security' more highly than the patriots.

Some Concluding Comments

There seems to have been an upsurge in nationalistic impulses and strivings, which is reflected in contemporary wars and related conflicts. I have argued that individual aggressive impulses, whatever their source, have relatively little to do with hawkish, militant values and policies, and that the nature of one's relationship towards one's nation is a much more potent factor. That relationship entails a complex set of feelings, cognitions and attitudes which must be distinguished in order to determine their implications for militarism and war. Both theory and the results of factor analytic procedures indicate the necessity to distinguish between at least two components of national feeling — one which we have labelled 'patriotism', and the other 'nationalism'.

A variety of studies have been reviewed that reflect important differences in the antecedents and consequences of patriotism and nationalism, despite the correlation that obtains between them. These data indicate that it is the nationalism component, in particular, that is significantly related to values and attitudes that are consistent with the readiness to support militant, hawkish policies. The patriotism component shares some of the attributes of the nationalism component, but the patriot is less militaristic. While he or she values the need for a military defence of the nation, there is no relationship between patriotism and support of national power and dominance. The difference between a patriotic and nationalistic identification is clearly conveyed in an analysis of predictors of these dimensions carried out in connection with the study of values and other attributes. When nationalism was controlled, the strongest predictors of patriotism were early father attachment and degree of realization of freedom of speech and press. When patriotism was controlled, the strongest predictors of nationalism were realization of economic opportunity and national security.

The particular value and socialization differences found between the correlates of patriotism and nationalism may be restricted to the United States and to values that US citizens hold to be important. However, we would anticipate that studies carried out in other nations would yield comparable differences, although not necessarily the same as those obtained with the US samples. We have already alluded to the findings obtained with a Japanese sample which replicate the factor structure found for the US population and, in addition, the correlation of patriotism with early father attachment (for the Japanese sample, patriotism was also correlated with early maternal attachment).

Assuming that similar findings are obtained in a variety of different national and ethnic settings, what implications can be drawn for public policy and future research? One might take the position that while the patriotism component has consequences different from those of the nationalism component, in the real world they are correlated and interwoven. Consequently, one might argue that we must strive to reduce national attachments, whether reflected in patriotism or nationalism. I do not believe this to be a tenable position. In my judgement, national and ethnic attachments are inevitable for most individuals. Quite apart from socio-biological arguments, the socialization process will result in national and ethnic identifications. It may be possible through particular socialization practices to 'undo' or reduce these identifications but one would be struggling against a very powerful mechanism. Moreover, there are a number of values and attributes associated with the patriotism component that have positive consequences for a society, and that one would wish to encourage.

In view of the data indicating that there are many individuals who are highly patriotic but who are not nationalistic, and who do not support 'hawkish', militarily aggressive policies, a more feasible policy is to utilize education and social action to sharpen the difference between patriotic and nationalistic attitudes and values. It is important that it be recognized and accepted that one can be strongly attached to one's nation without believing in national superiority or supporting national dominance. The difficulty of such an undertaking should not be minimized.

From the standpoint of national policy and international relations, the difference between national defence and national power and dominance is frequently obscured. Turks, Iranians and Iraqis may perceive the efforts of the Kurds to establish political autonomy as a threat to the integrity of their nations, and patriots as well as nationalists may share this view. Can this conflict between the Kurds and the nations in which they reside, a conflict which is replicated in many parts of the globe, be resolved without war? Perhaps, if the legitimacy and potency of the patriotic feelings of the different protagonists are recognized and understood: there are contemporary examples in which some such process has taken place. Russia and the Ukraine are not at war with each other; the Israelis and the Palestinians appear to be working toward a peaceful resolution of what had appeared to be an unresolvable conflict. I am aware, of course, that military and economic factors have played a major role in these peaceful outcomes. Still, these outcomes, while opposed by some nationals, appear to be perceived as compatible with the patriotic values and sentiments of most of the nationals in these respective groups.

The international community (United Nations) has already agreed upon the unacceptibility of national expansion through military aggression. It now needs to address the thorny problem posed by different ethnic groups as they seek independence, and establish guidelines that can be used to influence national and international policy in these situations. The international community can, at the same time, encourage the implementation of educational programmes that foster both patriotic values and international tolerance. Nirvana is not at hand. The next generation will undoubtedly experience conflicts and wars. However, war is not an inevitable consequence of the human condition. The kinds of analyses presented here should help foster policies that will reduce the incidence of conflicts and war to be experienced by the next generation.

12. War and Religion: An Unholy Alliance?

Helen Watson

As an anthropologist I feel rather ill at ease when faced with a massive subject like war and religion. General observations about the apparent acceptability of violence are complicated by differences in specific cultural and personal factors. Aggression and violence have different cross-cultural evaluations, and the perception of violence as a legitimate way of resolving disputes may vary within a society according to class, gender or age. A common strand in different analyses of war and violence is the concept of war as an institution or system. From this starting point my approach is to examine specific cases that illustrate how religious belief acts as a fuel that sustains violence.

The relationship between religious belief and action is a long-established preoccupation of social theorists. Durkheim, Weber and Marx respectively examined the structural, ideational and ideological dimensions of religion in relation to individual and social processes. I have chosen to examine the level of individual belief and its relationship to war as a way of showing how religious conviction affects experiences and perceptions of violence. The examples I will present derive from research in the Arab-Muslim world and Ireland. The views of war and religion to be discussed present individual explanations of why violence is acceptable from the perspective of men who describe themselves as combatants, defenders of the faith and warriors for God. I have used the term 'fighter' in my commentary on the narratives as a way of avoiding value-laden labels, since it is a commonplace observation that one person's freedom fighter is another's terrorist. The first-person accounts of the experience of war will establish a foundation for more general observations about the impact of fundamentalist religious faith and violence, and the part religion plays in fuelling the institution of war.

The fighters' narratives explain individual action in terms of religious belief in three different religious, social and political contexts. The three cases also differ in relation to time-frame in that they represent past, present and expected future conflicts. Despite the diversity of settings and religious affiliations represented, there is a high degree of consensus in the narratives. Religious belief is presented as a central influence on the fighters' behaviour and motivation. Religious identity influences the perception of the purpose and the goal of war. Religious doctrine, symbolism and imagery justify the fighter's actions and legitimize the war itself. Religious conviction sustains the fighter before, during and after the experience of violence. In each respect, religion provides a comprehensive framework of support for the individual which has psychological, political, practical and symbolic significance for our understanding of the institution of war.

The Explanation and Legitimation of Violence

The notion of a 'just war' has been examined within the Christian tradition. Santoni (1991) has traced the Christian history of the 'just war' to St Augustine's attempts to provide moral justifications for Christian participation in war. However, there are parallel concepts of sacrifice, struggle and divinely sanctioned conflict in other world religions which make violence an acceptable or obligatory response for members of the community of the faithful. The narratives will show that Christianity has no monopoly on the concepts of the 'good fight' and the 'just war'. All world religions have the potential to reinforce the conviction that a fighter is 'fighting the good fight with God on his side.'

The narratives of war and religion combine expressions of religious ideals and cultural values with secular notions of patriotism, nationalism and territoriality. The consequence is an explanation and justification of violence made up of many conceptual layers. In this respect it is important to recognize that religious belief is one component part of the individual's cultural world view, which itself is affected by the wider global context. The analytical issue is not whether the secular elements can be distinguished or split off from the religious ones. The force of 'belief' which sustains the notion of a 'just war' derives from the interconnected nature of the sacred and the secular.

It is possible to argue that a distinction between sacred and secular is problematic even within non-theocratic religions which posit the separation of the spiritual and the temporal, church and state. What can be called a 'sacralizing' process seems to operate in the narratives of religion and violence. The fighter's language blurs the boundaries of the sacred-secular and religion-politics by describing and locating temporal concerns within a religious framework. For example, a Protestant paramilitary fighter discusses his experience of murder via a series of metaphors and hybrid quotations which mix biblical and political rhetoric: 'the sword and the pen', 'the Bible and the bullet', 'he who is not with us is against us', 'not peace but a sword'. In ideological terms the general effect of the process is to transform a secular concern, such as ethnic identity or territorial sovereignty, into a quasi-religious article of faith. This 'sacralization' results in the expression of a political goal as the functional equivalent of religious dogma, which inspires action and provides a *raison d'être* for violence.

The Model of Suffering and Sacrifice

Religion plays an important part in relation to personal comprehension and public expression of the human costs of war such as loss, defeat, suffering and pain. Folk memories of violence and conflict are inherited by subsequent generations via narratives of the past. These contested histories detail the sacrifice and pain suffered by each generation's ancestors for the sake of the future generations. The Western view of war as the 'motor of history' seems to apply equally well outside the Western cultural context.

Religious imagery dominates in the following personal recollection of being a 'freedom fighter' for Israel. The speaker is a member of Gush Emmunim (the Bloc of the Faithful) an extreme right-wing Israeli group. Members of Gush Emmunim believe specific territories mentioned in the scriptures belong to the Jewish people by divine right. In an ironic echo of the Ayatollah Khomeini's axiom that 'politics is religion', it has been argued that its members hold the view that religion is the most perfect and exalted type of politics (Aran, in Marty and Appleby 1992). I met the speaker in the late 1980s when he was visiting Egypt with his teenage daughter. The visit was an 'educational experience' for the daughter, intended to show the base nature of the Arabs in their own social and religious environment. The following extracts give an impression of how religious and political notions are combined in a description of the 'struggle for our land'.

> I have been a fighter since 1947 and an underground activist. I was not afraid to die because I was playing a part in a struggle for

> my people. I fought as a young man for our holy places, the territory which God has set aside for His People, the Jews. In those days we were a few dozen fighters depending on His protection without military back-up. We were inspired by our war to win the place King David had chosen as the centre of his Kingdom before conquering Jerusalem,[1] so how could we fail to fight … to the death if need be?
>
> Young people these days are becoming weak and forgetful of their fathers' sacrifices. They think that the centuries of Jewish suffering have been eased by the reward of our State of Israel, but it is no reward. It is a small sign of what we have won by blood sacrifice, bravery and hard work which can never be permitted to be taken away from us. I will always be a warrior, this is the nature of a real Jew — the soldier is always under the skin … my battle fatigues are there under my cotton trousers. … My organization is the finest example of our people's struggle and heroic resistence. The best example is the ancient old struggle for Eretz Israel in Samaria and Judea.[2] Our goal is simple and absolute, it is set out in our scriptures, we must possess the land and conquer it. The problems of the violence of the enemy and our war against them are an equally straightforward matter, the Arabs want our land. … But there can be no debate about the word of God [for God] defines the land as ours, His gift to us. … Today's freedom fighters tend to forget the essential religious principles involved and put their faith in political arguments about states and international boundaries. Our right is historically documented in the scriptures since the days of Abraham over five thousand years ago. We have made the land fertile and followed the Mitzvah, God's commandment to be fruitful and multiply. If the next generations forget the sacrifices we have made, we have fought and died for nothing.

Recollections of the suffering of war have a powerful impact on those who played no direct role in the conflict. A feeling of obligation for a sacrifice which cannot be repaid creates enduring bonds between fighters and postwar generations. Arguably, the feelings of obligation, loyalty and inherited myths of intra-group suffering of the next generation will help sustain the ideology of the original enemy's otherness and difference. The result is a replication of the myths of heroes and villains across generations. Hobsbawm (1992) describes how, in periods of rapid change and political disorder, 'retrospective mythologies' of ethno-national difference can provide a potent 'call to arms'. If the representation of suffering/sacrifice in nationalist historiography is reinforced by religious imagery, the result is a double-edged literal demonization of the historic enemy.

Demonization and Holy Violence

Armstrong (1988) argues that the conflict in the Middle East cannot be understood in isolation from 'the murderous triangle of hatred' among Judaism, Christianity and Islam, the three religions of Abraham. She argues that a 'triple vision' is needed to view the three sides of the question which have been tied into historical relationships between the Christian West, Judaism and Islam from the time of the Crusades. Webster (1990) attempts to put this 'triple vision' into practice in his exploration of the blasphemy in the Judaeo-Christian tradition. His analysis of the kinship between anti-Islamic and anti-Semitic traditions provides evidence of the potent force of religious imagery when it is manipulated for political ends.

The Christian basis of anti-Semitism is established in the Gospel of John where it is reported that Christ pronounced the Jews 'children of the Devil'. Luther contributed greatly to the demonization of Jews by instigating the anti-Semitic motif, *Judensau* (Jew's Sow), having identified the Talmud with a sow. Webster (1990: 35) demonstrates the enduring influence of this image. He quotes a contemporary of Luther at the University of Wittenberg who elaborated on the *Judensau* theme:

> The Jews had been called and elected to be God's mouth ... but they closed their mouths to the flow of all good from God and opened their mouths to the Devil who filled them with lies, impiety, blasphemy. ... The rabbis, enemies of God ... do not understand anything divine. Instead of the flowing water of eternal life they suck the milk of a sow ... they eat nothing but excrement and dung. They take all their mysteries from the piggish Talmud, they suck all the impurity from the teats of swine. ... Having deserted Christ they adhere to a sow.

The powerful juxtaposition of images involving a kind of metaphorical pollution (the Jew with the unclean pig of the Jewish tradition) remained part of Christian iconography for at least six centuries. The force of the image lay in how it dehumanized Jews and degraded

Judaism. It also has an apparent misogynist element which echoes how female healers were depicted as bestial witches in medieval church propaganda. Cohn (1976: 16) has commented that one of the strongest and self-perpetuating aspects of the Judaeo-Christian tradition is 'the urge to purify the world through the annihilation of some category of human beings imagined as agents of corruption and incarnations of evil.' The *Judensau* motif was to gain full social and political significance centuries later in the 1930s when adopted by the cartoonist Julius Streicher in the National Socialist newspaper, *Die Sturmer*. In one cartoon the pig was labelled 'Jew's Publishing House' and the caption reads, 'When the sow dies the litter perishes too.'

At the time of the OPEC crisis, as a new chapter in the conflict between East and West was being written, Said (1973) focused attention on a sea change in the popular cultural imagination of the West. He argued that anti-Semitism was being displaced by a revival of ancient anti-Islamic sentiments via the transference of a popular anti-Semitic stereotype from Jewish to Arab-Muslim targets. Said's later critiques of the ideological foundations of Western cultural dominance have implied the existence of deeply embedded processes at work in the unequal relationship between East and West (Said 1992). It is possible to place the 'Orientalists' he describes within the same cultural or politico-religious tradition as the medieval Christian polemicists who demonized Islam as a prelude to war at the time of the Crusades. Religious imagery and the manipulation of divine revelation played a central part in this process in the early days of Christendom. Pope Innocent III equated Muhammad with the apocalyptic beast of the book of Revelations. In sixteenth-century woodcuts, one of the earliest forms of mass propaganda, a triumphant Christ is depicted crushing the head of a Muslim 'Turk' underfoot. Throughout the sixteenth and seventeenth centuries the Muslim Turk was identified in Christian propaganda with Gog and Magog, the hosts of the Satan described in the book of Revelation (see, for example, Scribner 1981). Indeed Muslim and Jew often lay side by side under Christ's feet.

Webster (1990: 37) comments, 'The tendency of Christians to demonise their enemies meant that realistic fears of Islam were increasingly overlaid by demonological fantasies in which Muslims in general, and Mohammed in particular, were seen as satanic beings.' The violence and religious-racist stereotypes associated with the 'Rushdie affair' must be understood within this historical and cultural context.

On a more general level, a kind of 'holy violence' can emerge as a consequence of the clash between religious traditions, blasphemy, demonization and the manipulation of religious imagery for political ends. Secular war propaganda has effectively borrowed the demonizing/dehumanizing tendency from the religious iconographers. As in the case of the *Judensau* cartoon, the direct line of descent from the religious to the secular is often unbroken.

Religion, Identity and Legitimate Violence

Part of my research in Ireland has concerned the ideology of religious difference across the sectarian divide. I have focused on perceptions of revenge killing among Protestant paramilitary groups where religion appears to be the primary factor in selection of a victim. Religion is a basic feature of identity in Northern Ireland. For example, being able to identify the religious identity of a stranger is a prerequisite for social interaction (Burton 1978; Cairns 1987). Once identity is established by any one or more 'cues of telling' (name, educational background, residence) a stranger becomes identifiable as 'one of us' or 'one of them'. In public and private discourse there is a close affinity between religious identity and political loyalty, i.e. the Catholic-republican/Protestant-unionist divide. There are long-standing debates about the religious, political or economic roots of the conflict. I find the search for the fundamental root cause less useful than a recognition of the interplay of factors at different levels such as the public and private, and the local, national and global. With respect to religion I agree with O'Brien's ironic comment that if religion is a red herring in the conflict in Ireland, then it's a herring the size of a whale. Religious identity is a gloss for ethnic and political difference. Religion is used to justify sectarian ideologies of difference. Religious images and symbols are brought into the rhetoric of 'otherness'. Perceptions of religious difference fuel the pattern of conflict and play a key role in sustaining the conflict in both active and indirect ways.

Unlike militant republican groups, Protestant terrorists operate without a secular concept of a 'legitimate target'.[3] In general terms, Catholicism is presented as the enemy of the Protestant community with the result that in some sense all Catholics can be perceived to be a 'legitimate target'. The importance of religion as a factor of Protestant political identity is explained by Bruce (1987). He argues that the first step to understanding the political behaviour of the Protestants of Northern Ireland is to see that 'loyalism depends upon its religious base … beyond [it] no secure identity is available … [Their] problem is defining the enemy and maintaining a sense of what is distinctive about Northern Irish Protestants'. Themes of religion, identity and violence combine in the narrative of a member of one Protestant paramilitary organization:

> The killings are about our future. Violence is our only means of safeguarding the Protestant heritage. We are fighting to defend our people; innocent women and children are being butchered by republican murder squads. Violence is a terrible thing, but for me it's a matter of self defence and a necessary struggle against evil. We will not allow our people to cower before the forces of oppression of the Catholic Church. We will never despair. It's clear from our necessary sacrifices that we are strong and defiant.
>
> If you draw a line separating politics from religion you can't understand the nature of the conflict. We are not fighting about religion or about politics alone. We, the Protestant activists, are defending our people against the Roman Catholic Church. The Church is the enemy in that its priests are the backbone of republican terrorism. The Church sanctions terrorism, gunmen are given the last rights, murderers are not excommunicated.[4] We have no quarrel with ordinary Catholics but they have to condemn the murders of their fellow believers. The problem is that there is no such thing as an ordinary Catholic: because of their religion they are in league with the republican movement. A Catholic will sympathize with the aims of the IRA whether they support it directly or indirectly. You might exclude one or two individuals from this rule but these would be individual Catholics you know and trust. I wouldn't go further than this; it's hard to say if any could be trusted.
>
> Murders and revenge are a way of life now. Politicians are spineless, so forget ordinary politics. The problem is religion, not religion with a capital R but the Catholic religion, which became corrupted centuries ago. In Northern Ireland the Protestant people know that if you scratch any Catholic you'll find a terrorist sympathizer under the surface; a gunman or a would-be bomber committed to the destruction of our union with Britain and the establishment of a Catholic state. Our Protestant people have a long history of defending freedom and fighting for justice and liberty. We fought for liberty against the armies of James II and Roman rule, and we defended democracy against the Kaiser and

Hitler. We are still fighting for God, Ulster and the monarchy as part of this tradition. We cannot let our people be killed without seeking revenge.

But we are shifting into a new tactical position now, new battle lines have been drawn. Too many have died so far to give up the fight. When we seize the initiative and strike the first blow we do so with the knowledge that we have become the legitimate aggressors. Our motto is 'an eye for an eye' and it is a statement of intent which will take us forward on the bloody path to our goal of a just peace.

The speaker uses religion as a way of constructing an unambiguous enemy who is depersonalized and to some extent demonized by affiliation to the perceived evil agenda of Catholicism. A focus on religion allows him to construct a depersonalized image of the victim of Protestant violence. The explanation of 'acceptable and necessary violence' rests on a view of the world as composed of rival camps representing the forces of liberty and oppression, good and evil. The rationale for violence depends on fusing and confusing the categories Catholicism, anti-unionism and terrorism to produce the enemy category of 'Catholic Republican Terrorist', the common enemy of the Protestant community. On another level, the rhetoric blends notions of justifiable defensive violence with that of the necessary evil of the pre-emptive strike which will win a just peace. The argument is that violence is acceptable because, in the words of the speaker, 'It is our only way of showing our enemies that we have unswerving loyalty to all that is precious to us … our heritage, our faith and identity, our love of liberty and belief in freedom.'[5]

The Duty of War

The narrative, which expresses religious motivation to fight in the future defence of religious belief, comes from a teenage Muslim boy from a poor district of Cairo. He is a member of the Muslim Brotherhood and favours the creation of an Islamic system of government in Egypt. He is talking about his conviction that there must be an armed rising of the pious against the corrupt and immoral sectors of society, explaining that he will fight in this war as a fundamental aspect of his religious and social duty.

Ruhollah Khomeini tells us that politics is religion. I also believe that faith is empty if one is not prepared to fight to achieve the best and truest form of religion possible for a good and just society. The evil in the modern world is a result of weakness and the love of luxury. It is not a part of true Islam, it has spead from the outside world and tainted our leaders, who in turn lead the people into corrupt ways. I weep with anger to think of how the actions of the Prophet and the martyrs of Islam have been undermined by the emergence of evil, corruption and greed in the society they created. The battle between good and evil will never cease, but the balance shifts through time as righteous men rise up to inspire the faithful to proclaim God's greatness by their actions. Force and violent struggle are needed, not just words. ... The time for struggle is near and I will sacrifice myself in the struggle against evil. I will kill my own brother with the same force as I would any non-believer who opposes the war against evil. The examples of the glorious Islamic past teach us that a concern for personal triumph and glory is not a just reason for battle. Even in defeat and suffering, it is the war which is righteous and the struggle which is just. ... The greatness and pride of the Islamic revolution came about because of the people's collective faith and action in the establishment of a true Islamic society. The result depended on millions of unknown heroes who set the pillars in place with blood-stained hands. Their sincerity was expressed in the spilled blood of their sacrifices for Islam and the future of community of believers. My belief is my life. My faith is my duty. In the fight ahead my blood will be proof of my faith in the one true God.

Fundamentalism and Fighting

In the narratives of war and religion presented, each speaker explains religious belief as a factor that motivates, sustains and legitimizes violence. Individual accounts of experience describe 'acceptable violence' in relation to the sense of participation in a 'just' or 'holy' war. In a general sense, the fighters are presenting a fundamentalist conception of religion. Fundamentalism is a term that tends to be associated with Islam in popular Western consciousness. 'Muslim' and 'fundamentalist' tend to roll off the tongue as if hyphenated, in much the same way as one talks about 'human nature'. However, there is increasing recognition of fundamentalism as a contemporary global trend which is proliferating in all world religions.

Marty and Appleby (1992) provide a comprehensive survey of fundamentalism in a wide range of social, political and cultural environments. Given our concern with religion and war, it is interesting to note that fundamentalism can be defined in relation to a single word: fighting. Fundamentalism is about fighting in terms of religiously derived goals, motivations, rules and weapons. The range of fundamentalisms discussed by Marty and Appleby is impressive, though worrying in what might have been predicted to be a secularizing, post-modern, *fin de siècle* world. A brief survey of the recent research will illustrate some significant militant aspects of contemporary religious fundamentalisms.

The emergence of fundamentalist religious movements in India is strong evidence against an assumption that official state secularism leads to the 'privatization' or decline of politicized religion. Western dichotomies such as religious–secular or sacred–profane are alien to India's major religious traditions. The double-edged sword of Sikhism symbolizes spiritual and temporal authority, representing the inseparability of religion and politics. The image of the sword is used by Sikh fundamentalists to express the Janus-like face of religio-political fundamentalism (Madan 1992). The violent potential of fundamentalism in Hinduism has been described in terms of a new socio-religious order which sustains the vitality of Hinduism in the modern world. Gold (1992: 581) asserts that fundamentalisms appeal 'to what may be the urban Hindu's lowest common religious denominator: a Hindu identity'.

Fundamentalism is often seen as a by-product of the impact of social and political change on national, cultural and personal identity. Fundamentalist movements in Theravada Buddhism in Sri Lanka and mainland South-East Asia emerged from the ashes of the classical religions and the cultural syntheses produced by colonialism and westernization. In a similar way, fundamentalism in Islam has been described as 'a reservoir for cultural and social turbulence' and a 'ready lexicon of protest and of promise' (Nash 1992: 734).

One of the most important Islamic movements of South Asia, *Jamat-Islami*, has the aim of 'resacralizating' social and political life. The movement's practical and ideological struggle represents 'an attempt to find a place in the new social arrangements' by 'reappropriating the processes of modernization' (Ahmad 1992: 510). Processes of Islamization within Shi'ism in Iran, Iraq and Lebanon have been described as a form of political assertiveness 'which does not entail loss of identity'. In this context fundamentalism 'provides renewed confidence in encountering the modern world with its new subtle forms of colonization' (Sachedina 1992: 450).

One common ideational characteristic of fundamentalist movements is the tendency to make the cognitive dimension of religion 'foundational and determinative'. In the case of Roman Catholic fundamentalism in North America, the priority is to determine correct belief as a model or charter for action. In this respect 'religion is based on a standardized, objective knowledge of God' (Dinges and Hitchcock 1992: 81).

The Haredim, a radical segment of 'orthodox' Judaism, embrace an activist traditionalism which depends on the manufacture of authoritative versions of history. The Haredim have 'a commitment to a particular constructed image of an idealized past ... The *Kulturkampf* which fundamentalism represents is therefore a struggle not only over what must be done in the future but over what was done in the past' (Heilman and Friedman 1992: 257). As in the case of Gush Emmunim, the interplay of the sacred and secular means that Jewish-Zionist fundamentalism is an example of both 'political religion' and 'religious politics' (Aran 1992). Although this wordplay implies that it is possible to label isolated elements of belief and action in an unproblematic way, perhaps the point is that 'religious politics' and 'political religion' have an identical potential in relation to the institution of war.

The first-person narratives have given an impression of religion as a fuel in the machine of war. From the perspective of the fighters discussed, the political and the religious are inextricably linked and presented as crucially interrelated elements. Fighting for the 'just cause' of a new socio-political order is defined in relation to religious belief, personal identity and political conviction.

The narratives have presented a way of generalizing about the impact of different aspects of religion on the creation, maintenance and validation of patterns of conflict. In exploring the relationship between belief and action, it appears that violence is acceptable when it is represented and understood as an essential aspect of religious duty. The expression of violence in a religious idiom legitimizes war and shifts it from the realm of the 'ordinary' on to a higher moral plane of existence. Religious justification also can negate the ambiguous feelings, contradictions and ethical considerations the individual faces in war. When a past struggle is presented in mythical form, defined by ancient religious law and overlaid by religious symbolism, it provides a potent model for future generations. Religious imagery can also be used to justify violence and 'decriminalize' the fighter by dehumanizing the enemy. At a fundamental level, all world religions provide believers with images of struggle, holy wars, sacrifice and the opposing forces of good and evil, which can create validations for action among believers.

If I have concentrated on cases which in effect 'demonize' religion as the 'motor of war', I will somewhat qualify my position by the following example of Gandhi. When asked about his apparent antipathy to Christianity, Gandhi replied that he had no reason to oppose Christian teachings; his quarrel was with most of the Christians he had encountered. On a parallel track, my argument is not that religion makes violence an acceptable way of resolving conflict, but that believers justify their actions as part of some morally superior, transcendent plan. In this respect, religious convictions obscure issues of individual and collective responsibility, which cannot but perpetuate the cyclical pattern of conflict.

Notes

1. Hebron.
2. Samaria and Judea are scriptural names for parts of the West Bank to the north and south of Jerusalem.
3. A 'legitimate target' was defined by a republican activist as 'a uniformed representative of the armed forces of the British Crown'. This notion of 'an unambiguous enemy' includes soldiers of the British army and members of the indigenous police and security forces, as well as any individual who is seen to represent, support or maintain British interests in Northern Ireland.
4. Paisley provides a characteristically clear-cut interpretation of the problem in religious terms which gives an unambiguous identification of the common enemy: 'Through popery the devil has shut up the way to our inheritance. Priestcraft, superstition and papalism with their attendant vices of murder, theft, immorality, lust and incest block the way to the land of gospel liberty.'
5. The speaker referred me to the sermons of Thomas Drew, a late-nineteenth-century preacher and 'denouncer of popery'. Drew is considered to have expressed 'the essential nature of the Protestant struggle for liberty from oppression'. The following extracts are from a famous sermon of 1857 which, according to Boyd (1987), marked the beginning of sectarian riots in Belfast. Drew addressed the audience with Matthew 5: 13–16: 'Ye are the salt of the earth. Ye are the light of the world. Let your light so shine before men that they see your good works and glorify your father which is in heaven …' The sermon provoked violent clashes against Catholic districts initiated by members of the congregation.

   'The Sermon on the Mount is an everlasting rebuke to all intolerance. Of old time, lords of high degree, with their own hands, strained on the helpless rack the limbs of the delicate Protestant women, prelates dabbled in the gore of their victims. The cells of the Pope's prisons were paved with the calcinated bones of men and cemented with human gore … The Word of God makes all plain; puts to eternal shame the practices of persecutors and stigmatises with enduring reprobation the arrrogant pretences of Popes and the outrageous dogmata of their blood-stained religion …'

He concluded with the account of the experiences of Protestants during the reign of James II: 'Then did Protestants feel intensely their oppression. Then were the Protestant churches nailed up and their congregations ordered to forgo their assemblies. Then were the lives and property of the Protestants a prey to despoiling priests. Then … the Deliverer landed at Carrickfergus.'

William of Orange, the 'deliverer' defeated the Stuart armies of James II in 1690. 'King Billy' remained the hero of loyalist Protestants. The Protestant Orange Order organizes an annual parade in commemoration of the glorious revolution' and victory over Catholicism.

Part III

SUB-INSTITUTIONS

Introduction

The chapters in this section are concerned with longer-term issues conducive to the maintenance of the institution of war. Three of these are selected — political history, economics and the arms trade.

Although many of the contributions to earlier sections stressed the importance of historical factors, the emphasis was on individuals' perceptions of the past, often deliberately distorted by government-controlled media. In writing about Japanese militarism, McDermott (Chapter 13) is concerned with the supposed facts of interstate relations. While acknowledging that the 'emperor ideology', the power of the military elite, and the nature of Japan's internal political economy may have had a role, he stresses the failure of Japan to establish adequate relations with neighbouring governments. Although the two centuries of isolationism espoused by the Japanese may be a Far Eastern phenomenon, McDermott's analysis shows how the seeds of militarism may be unrecognizable by contemporaries and take many years to germinate.

Wars often concern access to resources or markets. Japan's militarism depended in part on its needs for natural resources and markets available only outside its borders. Though such needs are not always the most immediate issues (for example, Drucker-Brown, Chapter 3), they are often there in the background. Neild, in Chapter 14, takes up the economic bases of war as a general issue. While believing that there is no immedate danger of our running out of primary resources and that technical fixes can persist almost indefinitely (views not generally held — see, for example, Smith et al. 1994), there are real dangers of economic conflict over natural resources leading to war, especially if there is a fundamental structural inequality between the parties involved. Neild discusses how the bases of conflict have changed from competition for empire before the Second World War, and how the emancipation and increasing military power of Third World countries have changed the pattern of conflict over resources. And even where conflicts seem to be primarily religious or ethnic in nature, economic differences between the conflicting parties may produce a secondary range of motivating factors.

To consider war as an institution with constituent roles of soldiers, munition workers and so on, as we have done so far, is of course a simplification. It must rather be seen as an institutionalized set of nested sub-institutions. Eisenhower (1961) aptly referred to it as the military-industrial complex, though perhaps the military-industrial-scientific complex would be more appropriate. Each segment of this complex is maintained by its own forces, but undoubtedly the major factors are the career ambitions and inertia of the individuals involved, and the bureaucracy of the organizations. The last two chapters take up particular aspects of arms manufacture and supply. Elworthy (Chapter 15) shows how the bureaucratic nature of the system of arms procurement in the UK leads not only to inefficiency and long delays, but to scientific research creating a 'need' for further weapons. Brzoska (Chapter 16) gives an overall view of the arms trade, a trade which is itself self-stimulating, a cause of wars, and a factor in the maintenance of the institution of war. There are, of course, complex issues here of both an ethical (Williamson 1992) and political-economic (for example, Elworthy and Ingram 1992) nature. For instance short-term gains derived from the arms trade by industry (including the employment of workers) provide an inertia to the system which even evidence that the trade has had an adverse effect on the economy of the exporting country cannot overcome. And the rapid increase in the military budgets has squeezed social services and economic growth in many countries of the south: poverty in the general population is no barrier to affluence in the military (Okoye 1994). Other aspects of the military-industrial-scientific complex are discussed in Prins (1983).

13. Modern Japan and War: A Problem with a Past

Joseph McDermott

Modern Japan has been forged in warfare. Between 1894 and 1945 it engaged in four major wars, lasting a total of 23 years and resulting in over 20 million casualties throughout Asia. Its armed forces during these years consumed over 40 per cent of its overall national budget and by 1945 had absorbed on a full- or part-time basis over half its adult male population. The Second World War ended not simply with some three million of these troops defeated and isolated on the Asian mainland, but also with the population of 66 Japanese cities heavily bombed, two of them unforgettably devastated by nuclear bombs (Dower 1990).

War and its needs have shaped the modern Japanese economy. Scholars have often noted the great stimulus European demands for Japanese goods had on Japan's growth during the First World War. In the view of Akira Iriye, it took this war to industrialize Japan. The Japanese military's influence on the Japanese economy was even greater. Between 1890 and 1919 military expenditure accounted for as much as 20 per cent of the nation's income. In the 1920s Japan's military budget admittedly dropped by 40 per cent, but the following decade and a half of war, beginning with the 'Manchuria Incident' of 1931, saw a return to normality: overall military expenditure consumed 30 to 50 per cent of each year's government budget. These years of war also saw a vast expansion of military production, in part because the wartime economy itself, during the 'Depression decade' of the 1930s, grew on average 5 per cent per year. From 1930 to 1944 the number of manufacturing and construction workers rose by two-thirds to nine and a half million, while heavy industry's share of the industrial work force rose from 27 per cent in 1930 to 68 per cent in 1942 (Iriye 1992: 23, 26, 50).

War also led to long-term changes in the workplace, as the government sought to transform the populace into a single family fighting for the nation. Between 1939 and 1943 the government exerted pressure on companies to establish two of the now characteristic features of 'the Japanese employment system', life-long employment and seniority-based pay. In the early 1940s it also extended social benefits to improve the health and morale of its soldiers and workforce, for example by establishing the present pension system for workers and a national medical insurance for over half the population. It is no wonder that John Dower has recently concluded, 'a great many of [Japan's present] institutions, ideas, practices — and leaders — were formed in the crucible of war' (Dower 1990: 66).

Defeat in 1945, a catastrophic defeat that resulted in the loss of over a quarter of the nation's wealth, led to the occupation forces' imposition of a post-war constitution rejecting all war by Japan. But that all too recent half-century of war remained a problem. The government, long dominated by packs of wartime politicians and bureaucrats resurrected by the Americans, slowly built up the military budget to become the second largest in the world today. These same government officials have also repeatedly fostered historical amnesia. Happily continuing the wartime distinction between the 'China Incident' and the Pacific war (the former was a term coined by the Japanese government to hide the reality of a protracted and bitter war, while the latter allowed the Japanese to discuss their military action in the Pacific as an anti-imperialist war), it attempted to replace 'aggression into China' with 'advances into China' in its school textbooks, and pinned the blame for the Pacific war on the Americans' refusal in the autumn of 1941 to sell Japan oil.

But some Japanese, including their wartime head Emperor Hirohito, have found it hard to forget. Tales of Japanese wartime abuse of Asian POWs are now being televised at peak viewing hours in Japan, and confessions of germ and chemical warfare are found in its newspapers and journals. Other Asian countries have proven even less forgetful. If only for their own domestic standing, their governments have protested against the Japanese tampering with the official accounts of the Second World War, and recently women forced to serve as Japanese army 'comfort women' have been airing their tales of grief throughout Asia in the hope of receiving a public apology and financial compensation before they die. The war, some five decades after it ended, is not yet dead for a country whose leaders now want others to accept its lead as a superpower.

Since the issue of modern Japan and war, like modern Germany and race, seems to have become a constant in twentieth-century politics and a fixture in its historical discourse, Japan's modern history is of singular import for any study of how war becomes institutionalized within the framework of a modern nation state. Scholars, mainly in Japan but also in the West, have not hesitated to provide a variety of explanations for this national and international disaster. First, and least relevant to us, is the claim, contrary to the wealth of evidence for Japan's previous three decades of militarism, that the 1931–45 period was an aberration, a disastrous but temporary deviation from the underlying drive in the economy and politics toward democracy and peace before and after these 15 years of war. A second approach has found the explanatory key in the strength of a pervasive military ethos. Taken in its crudest form, usually by less knowledgeable observers, this approach finds the culprit to be 'the old Samurai spirit and the dominance of the warrior caste'. Yet, from 1600 to the arrival of the US naval commando, Commodore Oliver Perry's so-called 'black ships', in 1853, at a time when European nations seldom let a year pass without engaging in a battle or war, Japan enjoyed the 'Great Peace'. The 2 per cent of the male population who ruled the country, the samurai, had in the early seventeenth century flatly rejected the adoption of Western guns, and by the early nineteeth century a sizeable portion of this political elite had turned into bureaucrats more gifted at swishing a brush than at swirling a sword. Meanwhile, commoners and aristocrats, the remaining 98 per cent of the population, had for the two and a half centuries of peace been strictly barred from any use or possession of arms.

More knowledgeable uses of the military ethos approach have stressed the late-nineteenth-century government's conscription of the peasantry, its efforts to indoctrinate the populace with 'the emperor ideology', and the lack of constitutional restraints on military activities. Over the first and second of these issues, the Japanese people are usually seen as innocent victims of state propaganda (despite the many links between this 'emperor ideology' and popular religion) (Gluck 1987). And with respect to the third, the military elite is seen as acting independently of all other elites. Beholden to the Diet only on budgetary matters, the military had direct access to their supreme commander, the emperor, who rarely objected to their wishes; the major opponents of military policy usually came from within the military, from those more aggressive than the military command. Criticism from the political and business elites was short-lived, because some feared assassination and others actually supported the military's adventures in China (Byas 1944; Beasley 1987: 36).

Some fruitful work has been done on linking the history of militarism with Japan's political economy. Rapidly being transformed from an agrarian to an industrial society, Japan, like Germany, is said to have retained much of its military past to support a threatened social order (Berghahn 1984). War, in this rather schematic explanation, is thus linked to a particular transition period, even though much of the social order of the 1930s and 1940s was freshly created to handle new conditions and problems. More useful is the emphasis on the business classes themselves. According to a view widely held in Japanese scholarly circles, the persistence of high tenancy rates in the countryside impoverished much of the potential domestic market and so prompted some Japanese soldiers and businessmen to seek out markets abroad. With the international economy often dominated by Western powers restricting access to the markets of their Asian colonies, the Japanese sought one-sided trade arrangements and the acquisition of essential raw materials. When this strategy confronted the opposition of Chinese nationalists, these businesses, we are told, patched up their differences with the military to support the warfare now needed to secure their markets. Informal imperialism gave way completely to formal imperialism, as economic and military elites sought a new political solution to their common problem (Beasley 1989: 7–19).

These three analyses of protracted Japanese militarism — the 'emperor ideology', the power of the military elite, and the character of Japan's political economy — all have something to offer any multi-causal analysis of why Japan engaged so often in warfare between 1895 and 1945. But they neglect the one causal aspect that has changed least since the 1930s and that could most easily lead to conflict again — the character of interstate relations in East Asia. Past advocates of this emphasis have commonly stressed the sense of injustice felt by Japanese in their dealings with the West. They write of the imperialists' threats to Japanese national security in the nineteenth century, the unequal treaties imposed in the nineteenth century, racial insults in the Treaty of Versailles negotiations, America's exclusive immigration policy, and its oil embargo. As all these themes portray the Japanese as the unfortunate victim, their tale of grievances has proved popular with Japanese nationalists and government figures. But such misfortune cannot adequately explain, let alone justify, the initiation of war, especially when the victims of Japanese aggression were not its 'Western aggressors'.

A structural analysis of the *longue durée* of Japanese interstate politics neither plays the moralist's game of identifying villains and victims, nor primarily tries to explain why particular states fight particular wars. Rather it seeks to understand why war repeatedly occurs in a given area. It focuses on long-term historical practices in the arrangement of states — the distribution of power, their manner of interaction, and their aims — in a potentially anarchic world. In seeing how their separate and collective policies have led them into conflict, at times breaking out into war, it pays close attention to the regulations of and threats to regional order and hierarchy. War, single or plural in occurrence, can signal such a fissure in the political hierarchy and thus become a means to overthrow a political hegemony.

This chapter's aims, then, are: (a) to specify the principles of the traditional East Asian interstate order, (b) to determine Japan's very limited political role in it, and (c) to assess the impact of such practices on Japan's acceptance of the treaty port system of the nineteenth century and its practice of imperialism and militarism from 1895 to 1945. Japan's relative position of power in East Asia has undergone sharp shifts of fortune. In the past it might have traded, sent emissaries and monks, and even paid tribute to China, or it might have done just two, one, or none of these. But even in the best of times it never established relations with its East Asian neighbours they all found acceptable, let alone agreeable. In modern times it is this failure to work with neighbouring governments — admittedly a difficult task in this past century of traumatic change and revolution — that has destabilized all Japan's efforts to find a place for itself in a new multi-state system of nation states in East Asia. While our focus will be on Japan, the same governmental disinterest in close or regular links with neighbouring states can be found deeply rooted in the ancient past and recent history of other East Asian states.

Japan's most persistent problem in its history of foreign relations might be said to have been China. Not that China ever attacked Japan (unless one counts the Mongol Yuan dynasty as Chinese) or that China directly caused Japan considerable trouble; the direct troubles were caused in fact by Japan. Rather it was the sheer scale of China's size, wealth, resources, and population that dwarfed Japan and the other East Asian nations and so made their dealing with the hegemony of China while retaining political autonomy so difficult. For virtually all of the two millenia up to the end of the nineteenth century East Asian politics rested on a truism of geopolitics: the ruler of China was the ruler of East Asia. Not only was China, the civilization, the ancient Greece and Rome for other East Asian states, but also China, the state, was usually the only superpower in their vicinity. Its army, though often slow to respond, was the largest and often the best armed force in East Asia (Fairbank 1968).

The resulting Chinese hegemony was evident in the unequal way the Chinese government interacted with other states, particularly in the Ming (1368–1644) and Qing (1644–1911) periods when past practices were codified into a comprehensive 'tribute system' of far greater complexity and universality than ever before. In theory, the states and peoples around China were to demonstrate their subordination to rule by the 'Son of Heaven', the Chinese emperor, by paying visits of tribute to the court. These visits were intended to inform tribute bearers of their duties and bring them trade profits. They travelled in China under official guard. They were often taken along roundabout roads to the capital, to keep them from learning transport routes and any strategic information. Maps and books on geography, military affairs, and the politics of the present and the preceding dynasties were never to enter their hands and leave the country (Miyazaki 1959: 232–54; Wills 1984: 5–25).

The hierarchical relationship that such a journey was meant to confirm was invariably dyadic and never multilateral. The Chinese emperor was almost always considered the superior party. The states that paid tribute to him and China were supposedly civilized by contact with his state and so obliged to submit obediently to his dictates. States like Korea and Japan, which adopted Chinese political institutions and culture, were regarded as more civilized than the North Asian peoples who retained more of their traditional culture, but never as equals to China.

Other states, of course, have attempted to establish such a structure of dominance over their neighbours, but few if any have maintained such a political world view and order for as long as China. If one wishes to make a comparison, most traditional Islamic states, for example, have based their world view on their religious convictions. The unity sought was not political as under the guidance of a Son of Heaven, but in Hourani's words 'essentially one of minds and hearts, not of political forms' (Hourani 1983: 19). Each state was far smaller than China and commonly regarded itself the sole legitimate ruler of Islam. Other Islamic states were considered 'established rebels' and non-Islamic states 'heathens'. Yet an Islamic ruler could and did engage in bilateral relations with all of these states, and even had contractual agreements with non-Islamic states (B. Lewis 1988).

Admittedly, the Chinese case was not as simple as the Chinese court liked to pretend. In reality, even during its heyday in the Ming and Qing dynasties, 'the Chinese world order was a unified concept only at the Chinese end and only on the normative level, as an ideal pattern' (Fairbank 1968: 12). The second millennium of Chinese hegemony, more than the first, saw many such departures from the ideal. Japan, Korea and Vietnam had by the end of the Tang dynasty (AD 617–906) completed their adaptation to basic Chinese institutions and practices, and over the course of the next millennium they developed more distinctive cultures of their own. More troublesome were the north-eastern and central Asian tribes which, having learned enough of Chinese technology and organization, proceeded to conquer and rule all of China for a third of the last millennium and much of north China for over half of that time (Beijing, it is seldom recognized, has been under Han Chinese rule for only a quarter of the last millennium). Yet control of the Chinese empire remained the key to East Asian power and order, and China saw no reason to change this system (Rossabi 1983).

Japan's role in this principal system of interstate relations in East Asia was minimal. So long as it could trade with China and its educated men study in temples and other centres of learning in China, its leaders saw little benefit in participating in a framework of interstate relations which relegated it and its divine emperor to an inferior position. But only on one occasion, in 1403, did it formally accept Chinese hegemony: its military government, for domestic political or fiscal reasons, accepted inferior status as 'king' to the Chinese 'emperor'. This act was soon nullified and ever after denounced as a disgraceful departure from proper rule. With Japan officially withdrawn from an active foreign policy in East Asia, it soon found its subjects, some supported by local Japanese rulers, joining with Chinese pirates to plunder the coastline of China (Kawazoe 1990).

The sixteenth century saw the potential for great change, with not just more frequent Japanese travel abroad but also the first arrival of Europeans in Japanese waters and politics. They introduced technology, Christianity and weaponry, but no new concepts and practices of international relations (which in fact would arise in Europe only a century or two later). But Japan's invasions of Korea in the 1590s proved disastrous, and its regular dealings with the Europeans came to an end in the

1630s when it barred all of them but the Dutch from its territory. Over two centuries — until Commodore Perry's 'black ships' arrived in 1853 — would pass before Japan had to give serious thought to foreign relations. Meanwhile, with its subjects barred from travelling abroad and virtually all foreigners excluded from its shores, Japan enjoyed its 'Great Peace' (Elisonas 1991).

Some scholars have recently argued that this view of Japanese isolation in the Tokugawa period (1600–1868) requires considerable qualification. Anxious to establish an independent Japanese strain of foreign affairs, they have unearthed a history of ongoing Japanese contact with other East Asian countries. Based on the presumption that Japan was the centre of the civilized world, Tokugawa foreign policy supposedly allowed for multi-state relations conducted with some flexibility. Korea was considered an equal but China and the Ryukyu Islands were inferiors (Toby 1984; Wakabayashi 1986). Chinese culture, from books to paintings, pervaded all strata of society, providing more information on the outside world than anything gleaned from Dutch writings. Economically, Japan engaged in vital trade with both East Asia and the West. For example, 'in the seventeenth and early eighteenth century Japan was the world's largest copper-producing country and one of its major exporters' (Jansen 1992: 16).

When examined critically, this native tradition of foreign policy adds up, I believe, to far less than its advocates wish. First, it conducted foreign relations only with emissaries and received them merely from Korea and the Ryukyu Islands, never China, Holland and the rest of the world. The Ryukyu in fact had been conquered by a Japanese *daimyo* (feudal lord) and its tribute was not that of an independent state. The sole case of importance thus is reduced to Korea, and that adds up to little. The equality between Japan and Korea can be traced not to a mutual decision to establish equal relations with one another but to their common (mis)treatment by the Chinese in the Ming dynasty's tribute system. Also, Korea sent just 11 embassies to Edo in the entire eighteenth century and stopped its relations with Japan in 1811. Its missions to the Japanese-controlled island of Tsushima may have continued up to the end of the Tokugawa period, but in the nineteenth century they numbered just eight, each lasting briefly and consisting of no trade. Overall, the Koreans remained so suspicious of a country that had invaded them twice in the 1590s that they never allowed Japanese emissaries beyond the port of Pusan (Kim 1980: 15–25).

Secondly, by no later than the early eighteenth century the Tokugawa government adopted an economic policy that paralleled its independent stance in foreign politics. It took economic autonomy as its overriding goal, and made import substitution its principal policy. Alarmed by a currency shortage attributed to the export of precious metals for foreign goods, it successfully introduced new technology for sugar, silk and other products, so that goods previously imported could be made at home. Such import substitution generally satisfied local demands, while enabling Japan to catch up to China's level of technology by the early nineteenth century and yet escape the South-East Asian trap of relying on the Chinese to introduce, operate and thus dominate the relatively advanced sectors of its economy. Yet this policy also meant it withdrew even more from the rest of the world. Even critics of the traditional view of 'secluded Japan' admit that the open practices they talk about had been curtailed, except for the import of Chinese books, through government order by the early nineteenth century (Daniels 1991: 69–70).

This type of foreign policy, political as well as economic, amounted to little more than a low-level copy of the Chinese world order, with markedly less success in attracting clients. Like China, Japan often sought to prescribe foreign affairs according to protocol and ritual, to assure submission to its political authority and eliminate the need for serious thought about other states. One fatal consequence was Japan's continued inexperience with alliances, friendly or otherwise. In fact, Tokugawa writings do not even consider the possibility of Japan benefiting from a relationship with another state. The Shogunate viewed the rest of the world with fear, choosing to live in a 'hall of mirrors', complimentary rather than complementary.

This tradition of very limited diplomatic relations with neighbouring states was essentially practised by all East Asian states, including China. All these states were indifferent to encouraging the development of distinct rules, practices (other than ritual) and institutions for regulating complex forms of diplomatic interaction. Such mutual isolation let them escape the centuries of constant warfare that scarred the relations between early modern European nations. But the gains in East Asia were largely temporary, as these states had not yet addressed, let alone solved, the basic problem of how they were to deal with one another on terms they all found acceptable.

This challenge could be ignored while other nations stayed away. But it came to the fore once the Western gunboats, abetted by this vacuum of state relations among China, Japan and Korea, 'opened' these countries up. With the old tribute system seriously weakened, the structure of foreign relations which might replace it became the West's most demanding and disruptive challenge to East Asia. The dilemma was only to be confounded by Japan's Asian misadventures over the past century.

Japan's first response originated in a fate it shared with China — the treaty port system established by the British and ultimately by other Western nations to secure from Japan and China certain 'unequal treaty' privileges under the rubric of free trade. From 1859 to 1889 Japan was forced to give up tariffs and to let Western merchants run businesses in a few coastal cities. At the same time Japan realized that it might use this treaty port system to extract similar political and economic privileges from its East Asian neighbours. In 1870 it sought in vain to secure unequal treaty rights from China, but won them in 1895 after its victory in the first Sino–Japanese war brought down the tribute system, which China had continued to use in its diplomacy with its other Asian neighbours. With Korea, success came sooner, as in 1868 Japan relegated Korea to an inferior place in its hierarchy of states, in 1876 it won certain trading rights from Korea, and between 1895 and 1905 it annexed Korea into the status of a colony (Beasley 1987: 43–4, 86–90). Invited in 1899 to attend the International Hague Conference on war, a Japanese diplomat expressed to his Western counterpart an understandably cynical assessment of his country's progress: 'We show ourselves at least your equals in scientific butchery, and at once we are admitted to your council tables as civilized men' (Best 1979: 23). Japanese interest in Western writings on international law, particularly those arguing for the equality of nations and the power of reason over force, proved shallow, since they saw Western nations pursuing self-interested policies at odds with their professions of belief in international codes based on natural law (Blacker 1964: 124–32).

Japan's successful use of the treaty port system, however, did not resolve the problem of what new system of international relations might be established with the approval of all East Asian nations. Its rise in political and military status, so hard to achieve under the Chinese tribute system, depended heavily on its links with the Western powers. Through an adroit use of Western technology and organizational forms it had increased its power far beyond any level China had traditionally attained from its vast territory and huge population. But, as its rise clearly came at the expense of China and Korea, its relationship with these East Asian states and peoples discernibly deteriorated. Moreover, whereas Western governments were interested principally in enabling their companies to acquire profits in the China market, Japan's goals were both economic and military. Originally, its leaders may have followed Western priorities. But, after their victory in the Russo–Japanese war of 1905, Japan's representatives north of the Great Wall increasingly pursued both economic and military goals on their own. In other words, after 1895 they sought not only an informal empire in China shared with the West but also a formal empire of their own in Korea and Manchuria, the former being a state that had previously accepted Chinese suzerainty and the latter being the sparsely settled homeland of the Qing dynasty's ruling elite (Beasley 1987: 64, 85–171, 163–9; Osterhammel 1986: 290–314).

By 1931 this pair of parallel Japanese foreign policies was in collision, and the Western powers concurred with China's long-held conviction that in fact these policies were in collusion. The Anglo-Japanese Alliance of 1902 (Japan's first military or political alliance in its entire history) had been terminated by the British, who wished to draw upon American military power to restrain Japanese expansion through tripartite agreements made in Washington in 1922 (Nish 1966). With the Japanese invasion of Manchuria in 1931, most Western nations concluded that Japan's military aims overrode its diplomatic stance. Yet, they merely called on the League of Nations to write a report about the Manchurian crisis. As their primary interests lay at home far away from Japan during these Depression years, their failure to act upon the League's eventual criticism of Japan's invasion involved surrendering whatever claims they had to secure an East Asian order. Japan nonetheless took offence at this faint criticism, withdrew from the League, dropped all the arms limitation treaties it had signed in the 1920s, and sought to practise a fully independent foreign policy for the first time in the modern period (Nish 1966 and 1993).

This Japanese pursuit of diplomatic autonomy occurred when both its military and its companies had attained a remarkable regional dominance, particularly in China. By 1931 its China investments and trade were the largest of any foreign nation's, as the China market was becoming for Japan what the Western powers had long hoped it would become for them. Thus Iriye (1992: 25, 35) writes:

> Japanese exports to China doubled after the war of 1894–95, tripled after the Russo–Japanese war, and quadrupled during the European war. And Japan's share in Chinese imports increased

> from less than 3 per cent before the Chinese–Japanese war to more than 10 per cent after the Russo–Japanese war and to around 30 per cent during the [First] World War. There were parallel increases in China's share in Japanese imports, but Japan continued to enjoy substantial surpluses in the bilateral trade. Because … Japan suffered from chronic trade deficits until 1914, especially with regard to Western nations, the favourable balance of trade vis-à-vis China was one bright spot in the picture. Conversely, for China this meant that a significant portion — sometimes as much as 40 per cent — of its overall trade deficit was accounted for by the Japanese trade. So in some sense China unwittingly assisted Japan's economic modernization.

Before the 1930s, Japan's militaristic 'search for autonomy' had often not been closely linked to its programme of achieving economic modernization in part from its China trade. But Japan's growing militarism and its vast state-sponsored investments in Manchurian industry in the 1930s deepened its commitments in China. The growth sector of its China exports switched from textiles to heavy machinery, as some Japanese industries saw their future tied up with the Japanese army (Mizoguchi 1989: 10–30; Beasley 1987: 7, 10). Whether this link-up was necessary or inevitable is dubious. But Japanese pursuit of it doomed all calls for pan-Asian autonomy and outraged the growing number of Chinese nationalists intent on restoring order to warlord-riven China and on reviving China's regional hegemony. Japanese officials frequently insisted that they could deal better with China directly rather than through international intermediaries. But such negotiations repeatedly broke down, to be followed by further acts of aggression which culminated in the outbreak of war with China in 1937. Having followed the West into China under the unequal treaties, Japan now sought political autonomy and military hegemony in East Asia just when Chinese nationalism made it most difficult.

Thus, in a modern regional order bereft of, firstly, a hierarchy of power recognized by all the concerned states and, secondly, commonly understood rules and practices for the interaction of these states, it was hard to speak meaningfully of an 'East Asian world order'. Japan's militarism, to the extent that it was rational, was a desperate and suicidal effort to fill that vacuum. Its failure led to a prolongation of Western dominance of East Asia, to the promise of a resurgence of Chinese hegemony, and tragically to a continuation of the structural dilemmas that have made Japan's peaceful entry into regional politics so difficult and its wars so common over the past century. What gives cause for concern today is that far less has been done to address this cause of Japan's militarism than all the others. With the end of the cold war and with Chinese and Korean suspicion over Japanese interest in using 'neo-Asianism' to forge close ties with the rest of Asia, a problem with a past may unfortunately have a future.

14. Economics and Conflict

Robert Neild

The relationship between economics and conflict is a murky subject. One cannot hope to find precise laws saying that in this or that state of economic affairs, somehow defined, international war or domestic conflict will or will not occur. Economic causes are always mixed with other causes of conflict, and anyway society is always changing. But that does not mean one can say nothing. One can look with an economist's eye at the phenomena that may provoke conflict, and one can look at history to see what patterns one can find. If one does so, it quickly becomes evident that it is with respect to the pursuit of natural resources that purely economic motives for war are most clearly identifiable in theory and have been most prominent in practice; and secondly one can see that the pattern of forces making for conflict over natural resources has been changing.

I therefore propose to concentrate on conflict over resources. I shall also consider very briefly what, if anything, one can say about the economic element in the causes of all those ethnic, religious and nationalistic conflicts that are now occurring within nation states — or often within what were nation states.

The Meaning of Resources

Resources in this context are the natural resources from which we derive by extraction or cultivation what the economist calls primary products, meaning raw materials, foodstuffs and fuels. Natural resources are of two kinds: deposits from which minerals and hydrocarbons are extracted by drilling and mining; and surface areas on which foodstuffs and industrial materials (for example, cotton, wool and timber) are produced by agriculture, fishing and forestry.

In using the term primary production the economist is making a threefold division of the economy into the primary sector, where the products of nature are extracted; the secondary (or manufacturing) sector where these are transformed by men and machines into finished physical products; and the tertiary sector, which produces services, with labour as the main input.

Fixed Supply and Fixed Location

Two features of primary production which distinguish it from secondary and tertiary production make it a potential cause of territorial conflict.

First, primary production can take place only where the necessary natural resources are located. Hence there is competition for the possession of territory of a kind that does not generally occur with respect to secondary or tertiary industry. For manufacturing can be undertaken at any place where machines, labour and knowledge of the techniques can be assembled — and the necessary primary products can be imported. And the same is true of services: haircuts or the services of a bank can be produced wherever labour with those skills can go or be trained.

Secondly, there is on the planet a fixed quantity of natural resources from which primary products can be derived. Because they are in fixed supply, natural resources have a scarcity value. Ownership of them will bring you a pure rent. If, as an owner of land or mineral deposits, you produce primary products yourself, for example as a farmer who owns his land, the rent will be part of your operating profit — unless you are on marginal land where you only just cover your costs. If you are an owner who leases land for cultivation, or leases a mining concession or a drilling concession to someone else, you will charge a rent or a royalty. The rent will be directly visible.

It is important to understand the nature of these rents. They derive solely from the scarcity of natural resources. They are pure windfalls. If you come into possession of a deposit of a scarce mineral or an oilfield, or a piece of land that is sought after for cultivation, it will yield rent according to the value of the primary production that can be got out of it. That rent will accrue to you even if you do nothing but employ lawyers and agents to lease concessions to others who will exploit the resource; it will accrue to you even if you sit back and behave as an idle landlord.

The magnitudes in question can be colossal. Think of the value of discovering and claiming a gold-field or a diamond deposit or an oilfield. Think of the great gold rushes; of Rhodes tearing into Africa; or of the oil fortunes that the Arabs derive from happening to be sitting on oil — or the oil fortune that Mr Gulbenkian derived from getting an oil concession. These windfalls are so great that the pursuit of them is bound to be intense.

Scarcity value does not generally attach itself to the possession of manufacturing facilities, since machines are reproducible: there is no limit to how many can be produced nor to how many people can enter manufacturing production (subject to costs of entry) — so long as primary products are available to be processed into final products. Services too are reproducible without limit.

Thus the pursuit of rents in fixed locations is a general and fundamental feature of the economics of primary production that we can identify as a potential cause of territorial conflict.

Cyclical Characteristics

There are other features of the economics of the primary sector that have a bearing on the possibilities of conflict — and on the fears of conflict that sometimes gain currency.

Prices of primary products are volatile unless regulated. If demand for a primary product changes, or the supply of it changes (for example, because of bad weather harming a crop), the result, in the absence of regulation, will be a change in price on a commodity market, which is an auction market where prices change all the time. By contrast, if demand for a manufactured product changes, the main effect will be a change in inventories and then a change in output by the producer; price-trimming will be a second-order phenomenon.

One result of the volatility of primary prices is that there have been big cycles in primary prices associated with changes in the general level of demand and employment in the industrialized countries. When the general level of demand rises in a boom to the point where there is pressure on supplies of primary products, there is typically a surge in primary prices. These surges have been particularly pronounced in wars, both because wars usually bring with them budget deficits and hence high demand, and because wars generate speculative scares that supplies will be interrupted.

There were great surges of this kind in the two world wars and during the Korean War; and there was another in the 1970s when the oil price was pushed up and other primary prices rose too, though less dramatically.

When these surges occur, there are cries that we are running out of natural resources, that the supply of primary products is drying up, that there will be conflict over them. But then the boom ends, primary prices collapse, the alarm is forgotten and is displaced by fuss about surpluses; and that continues until the next upswing. We have fairly recently been through a cycle like this in which, as you will remember, the Club of Rome produced a rather alarming report; and we went through just the same thing at the time of the Korean War.

There are several explanations for why the 'alarms' recede. The first is that when primary prices rise there is inflation, since primary prices feed through into the prices of almost everything; and the reaction in the industrial countries to inflation is that governments reduce the general level of demand and employment by tightening the budget or tightening money. There may be a delay until the end of a war or until there has been an election or until the economists invent a new story for why deflation will be painless and good for you — monetarism was the last story — but sooner or later demand is restricted. That reverses the process. It causes primary prices to collapse again and punctures the scare.

That is a short-term explanation. It might still be expected that in successive cycles of that kind primary prices, owing to the exhaustion of natural resources, would rise relative to the price of manufactures. But I know of no evidence that that is so — though measurement admittedly is not easy. One can only infer — tautologically perhaps, but tautologies are not useless — that forces operating on demand and supply in the market for primary products have been sufficient to prevent that happening. On the demand side, we can suggest that the tide of technical progress has been material-saving; and that savings, by means of less wastage and the substitution of alternative materials or fuels, have been induced where there has been a disproportionate change in the price of a primary product. Certainly we can observe a long-term down trend in the use of primary products per unit of output. As you would expect, products are becoming more refined and incorporate less material per unit of total value. On the supply side, we can suggest that price rises induce more exploration and development of primary production; and we can see that technical progress has caused amazing increases in agricultural yields and the development of all sorts of synthetic materials; and we know that technical progress continues apace.

There are so many variables and unknowns when it comes to predicting the course of primary prices that making economic models is not much use. One has to rely on judgement, for what it is worth.

There seem to be no better grounds now than there were at any earlier time in my life — and I am 69 — for saying that we are running out of primary products, let alone that the impending exhaustion of natural resources is in itself becoming a cause of conflict. Since Malthus in 1798 — nearly 200 years ago — published his *Essay on the Principle of Population as it affects the Future Improvement of Society*, arguing that the growth of population was tending to outstrip the produce of the land, resource pessimism has been recurrent, but the economic system, aided by the advance of science, has been remarkably successful in inducing, subject to cyclical fits and starts, changes in demand and supply that bring the two into balance. It is a process that is not likely to come to a halt suddenly. That we may irreparably damage the ecosystem as we pursue economic growth and exploit the natural resources of the earth is another matter; it is a question of daunting dimensions and difficulty, which is beyond the purview of this chapter.

Besides the competitive pursuit of profits from natural resources, there are two other motives for the pursuit of natural resources that are potential causes of conflict.

One is the argument that the importer needs an assured supply and for this purpose has an incentive to own or lease natural resources on such terms that he can be the producer and can control the volume and quantity of output. It is an argument that can hardly be distinguished from the pursuit of profits. For the motive is again connected with profit (the benefit to profit of an assured supply), and it is the totality of all the arguments that concern profit that will lead to the decision to engage in the exploitation of natural resources.

Secondly, there is the strategic argument that you must be sure a potential enemy cannot strangle you by cutting off your supplies of primary produce in the event of war. It is a classic argument for protection of domestic primary production, invoked most often with reference to food. As regards imports for which there is no possibility of domestic substitutes, it leads clearly to the argument that you should build up strategic stocks of a commodity about which you are worried, an action that will be more cost-effective the shorter is the possible war; it tells in favour of a strong navy if that will help you to keep supplies flowing — a point that cannot be taken for granted: you may be too small to sustain a large enough navy, or land-locked like Switzerland, or it may be that surface ships are becoming so vulnerable to missiles and submarines that navies are ineffective in protecting shipping; it tells in favour of being on friendly terms with the supplying countries so as to keep them on your side.

The Pattern of Potential Conflict

The risks of economic conflict over natural resources leading to war are greatest with respect to the 'developing countries' in the Third World. In commercial relations among the advanced industrial countries, there is bickering over support of agriculture and a few other commodities, for example, coal. But in the secondary and tertiary industries, which dominate the advanced economies, the network of reciprocal interests and common interests has been progressively intensified by the growth of trade based on cheap transport and communications, by the emergence of multinational firms and by the internationalization of capital markets. Moreover, after a long history of wars Europe has achieved a measure of political unification. The fact that in the EEC the clash of economic interest is sharpest in agriculture is perhaps because it is in agriculture that *national* ownership continues and the internationalization of ownership has not occurred. If you ask who has a producer's interest in the fate of Shell, the petroleum company, it is certainly not just the shareholders or employees in the Netherlands and Britain, the two countries where the company is formally based; it is people all over the world. But if you ask who has a producer's interest in French, British or Danish agriculture, the answer is French, British or Danish farmers.

On the other hand, the developing countries typically have low income (there are exceptions, for example the oil emirates) and are heavily dependent on their natural resources and on the advanced countries to exploit those resources and buy the produce.

The conflicts of interest that can occur over natural resources in the Third World can be put into three categories: conflicts between one or more advanced countries and a developing country over possession of rights to exploit the latter's natural resources; conflicts between rival advanced countries over the exploitation of the natural resources of a developing country; and conflicts between neighbouring developing countries over the possession of territory containing natural resources.

Let us look briefly at history to see what it tells us about the evolving incidence of the first two of these three types of conflict.

But first a summary of the argument so far. The main points are: the basic economic cause of potential conflict over natural resources is the pursuit of the rents peculiar to the ownership of natural resources. The pursuit of assured supplies has been another motive for seeking ownership of natural resources, and so has the strategic argument. Judged by the economics of the matter, the risk of conflict over natural resources is small between advanced countries where common interests in secondary and tertiary production and trade are dominant and a considerable degree of economic and political cooperation and integration has been achieved; it is greatest with respect to the Third World where secondary and tertiary industry is small, though growing, and primary production is still often dominant.

The Historical Pattern

The history of conflict over resources is at the core of the history of imperialism. But imperialism has not been just about the pursuit of natural resources. That may have been its focus in the earlier stages when it was concerned with command of the spice trade, with the pursuit of gold in Latin America or with the trade in slaves for plantations: when imperialism was predominantly mercantilist. But by the nineteenth century British imperialism was a process of 'integrating new regions into the expanding economy' through the extension of both formal empire, where Britain ruled, and informal empire, for example in Latin America, where commercial or strategic integration could be achieved without having to impose British rule. Both can be seen as 'political functions of the extending pattern of overseas trade, investment, migration and culture' (Gallagher 1982). Here we see what has been called the 'imperialism of free trade'. There was emphasis on opening markets for the manufactured products of the industrial revolution as well as on investment of surplus income and migration of surplus population to generate a flow of primary products in return. It is remarkable how closely this description fits the role of the United States in the world since the Second World War.

That this was how policy evolved does not mean that the pursuit of raw materials was not important in the nineteenth century and seen to be so. Its importance was perhaps emphasized more by Continental economists who jealously analysed what Britain, the dominant industrial and imperial power was doing, than it was by British economists. Certainly it was emphasized strongly by Friedrich List, the great German economist who argued in the 1830s that the key to economic development, discovered by the British, was to exchange manufactures for primary products. Britain, he argued, having acquired an empire from which to obtain primary products, was running ahead of everyone else and was pushing free trade as a means of conquering the market for manufactures of Germany and other potential rivals. He advocated protection of German industry and the acquisition of colonies, as well as investment in infrastructure and education, as the way to react to the British challenge and generate development of the German economy (List 1909).

Indeed the study of 'national economy', with its mercantilist emphasis on development of the nation, is an understandable preoccupation of economists in a nation that lagged behind in the race for development and strength. It stands in contrast to the preoccupation with free trade on the part of economists in Britain, the leading industrial nation which confidently, perhaps over-confidently, was enjoying a competitive advantage in manufactures.

The main conflicts over empire were between European nations, and they were mostly fought out on land in Europe or at sea, not in the imperial territories on distant continents. The victor in a European war made the vanquished yield colonies. In the imperial territories overseas, the indigenous peoples could usually be overcome pretty easily by brute force or threat of force, since they had few, if any, modern weapons — a condition that did not obtain, however, when it came to fighting European colonials in North America or South Africa.

Competition for a share in the pursuit of formal or informal empire has persisted for a long time. It was a cause of belligerence in both world wars: Germany before 1914 thirsted for colonies for reasons of status, economic gain and military adventure; Japan in 1940 was imperialistic, seeking raw materials; and Britain was preoccupied with the defence of its empire and the routes to it.

Since the Second World War, there have been two developments that have shattered that pattern of potential conflicts. The colonies of the old formal empires have achieved constitutional independence; they have sought to develop their economies and, as part of that process, they have commonly sought, by means of expropriation or taxation of foreign owners, to claim for themselves a greater share of the rent from their natural resources. This has meant that the advanced nations' conflicts of interest over natural resources in recent decades have been predominantly conflicts vis-à-vis autonomous Third World countries, whereas previously, when the competitive struggle for empire, formal and informal, was taking place, the conflicts occurred predominantly between advanced nations. The conflict with Mr Mossadeq when he nationalized the oil industry in Iran is an early example of the new pattern.

Secondly, the peoples of the Third World now have sufficient knowledge of modern weapons, and ease of access to them, for it to be impossible for the armies of the rich countries cheaply to occupy their countries and subdue them. The peoples of the Third World countries do not need to acquire the most expensive or complicated weapons to achieve their militaristic goals. Vietnam and Afghanistan illustrate the point — though they were not wars fought over natural resources, unless you invoke a domino link to resources lying in countries beyond the area of conflict.

These developments have had in their turn two consequences. In place of the use of old-fashioned colonial warfare in pursuit of imperial interests in the Third World, we witnessed during the cold war the growth of sub-war, meaning the use of military aid, military and civilian advisers, economic aid, bribery, covert operations and the promotion of civil war as a means of toppling regimes that threatened the perceived interests of a powerful advanced country and as a means of supporting regimes sympathetic to those interests. The main western combatant in this kind of sub-war was the United States. And there were no half measures about the way it was done. US government documents recently released tell us that in 1954 the CIA was called upon to be 'an aggressive covert psychological, political and paramilitary organization more effective, more unique and, if necessary, more ruthless than that of the enemy' (*International Herald Tribune*, 30 August 1993).

The problem in the present context is to say how far these sub-wars were pursued for the sake of economic (or strategic) interests in natural resources, as distinct from (a) fear that the Soviet Union or one of its allies, for example Cuba, might establish military bases; (b) ideology, meaning blanket hostility towards the rival faith as regards religion, human rights, economics and everything else, because it is a challenge to cherished beliefs and values; or (c) economic interests in assets other than natural resources.

I have no specialized knowledge of local conflicts, but my impression is that motives were very mixed, not to say confused. In Central America, I think it is true to say that the socialist governments of Cuba, Chile and Nicaragua, with which the United States has had sub-wars, had all expropriated United States firms that were exploiting natural resources. But all the other motives mentioned in the previous paragraph were in play too. Among these, strategic concern to sustain supplies of the relevant primary products seems not to have been acute, since the USA embargoed trade with the offending countries and has continued the embargoes for long periods against the offending governments. Ideology and economic interest seem clearly to have been at work, and mixed in with ideology was the fear that the Soviet Union was trying to get into those countries in a militarily threatening manner.

The second consequence of the constitutional independence of the countries of the Third World, together with their highly armed condition, is that if their neighbours possess rich natural resources they may try to grab them. As a result of the ending of traditional empire and the competitive arming of the Third World during the cold war, the third type of conflict listed above — conflict between neighbouring developing countries over the possession of territory containing resources — has been let loose. The Middle East, where oilfields of huge value lie close to frontiers in barren lands, is where you would expect to find trouble of this kind, and there indeed you have the example of Saddam Hussein's attempt to grab oilfields from Kuwait.

It is not at all self-evident that this kind of conflict poses much of an economic threat to those who purchase the produce of the land that changes hands. Consider the case of Saddam Hussein. He, like anyone else who grabs an oilfield, would have wanted to sell the oil to Europe, Japan and the United States. He, like the other Gulf states, has nothing else to sell; they all badly want income now that their appetites for Western goods have been whetted; and the present oil-importing countries are the only places to which they can sell. Nor, out of self-interest, would Saddam Hussein or anyone else who got control of substantial oilfields be likely to push the price very high. OPEC has already learned that such a policy becomes self-defeating since it reduces demand for oil by inducing fuel-saving, substitution of other fuels and oil exploration.

But this kind of economic reasoning is not a reliable guide to what may happen in the event of conflicts over natural resources between Third World countries. Local conflicts over resources bring with them the risk that the rich powerful countries which import oil and other primary products will wrongly interpret what happens locally as part of a geo-strategic plot. I do not mean that there have not been and will not be geo-strategic attempts to get a stranglehold on natural resources on which your opponent, or potential opponent, depends. That is done regularly in war; and in peacetime there are people in the intelligence and foreign policy communities whose job it is to watch over that kind of action and advise ministers as they frame foreign policy. Since each side in a political confrontation will assume that the other will be playing these games, and each has only limited and uncertain knowledge of what the other is doing and planning to do, the possibility of misinterpretation is substantial. Sometimes a threat of this kind may be underestimated; but the chances are no less that it will be exaggerated, and once exaggeration begins, it can rather easily run wild as analysts pick upon bits of evidence that fit their preconceptions and as politicians take up the alarm as a means of rallying the nation to arms.

During the cold war the view, which quite probably was right up to a point, that the communists were out to get a stranglehold on the natural resources on which the West depended, was a spasmodic theme in the strategic rhetoric of the West. The last spasm was as late as 1979/ 80. Fear of a resource war was one of the planks on which Mr Reagan came to power in 1980 promising rearmament, rather as Mr Kennedy came to power in 1960 playing on fear of an alleged missile gap and promising rearmament. Resources, said President Reagan, were running out; the Soviets were moving into the Gulf and Angola to threaten the strategic supplies of the West. But the resource war, like the missile gap, faded away. Mr Reagan soon forgot all about it. So much so that he set about trying to cut the United States stockpile of strategic materials in order to save money (Finlayson and Haglund 1987).

Ethnic, Religious and Nationalistic Conflict

The interpretation of the role of economic factors in the ethnic, religious and nationalistic conflicts we see around us today is a daunting task in comparison with the interpretation of international wars over resources. Economic grievances seem always to be part of the mutual hostility that drives ethnic, religious and nationalistic conflicts. But that does not mean that economic injustice between groups is on its own a *sufficient condition* to start a conflict: the reaction of the economically deprived can be apathy, if it is not accompanied by a sense of grievance against others on ethnic, religious and nationalistic grounds. Nor can one say that economic injustice is a *necessary condition* for the start of conflict. If you think of the hounding of heretics and the pursuit of holy wars, in ancient and modern form, you can scarcely escape the conclusion that religious differences alone can be sufficient to explain the onset of conflict. But what one can say with some confidence is that feelings of economic injustice, whether or not they were there at the start of a conflict, are important in the processes by which it develops and gathers intensity. In other words, economic conditions, and people's perceptions of them, are important in the dynamics of conflict. For example, if there are ethnic or religious differences between two groups, economic differences they would otherwise accept in the conditions in which they live may become a cause of resentment and so help start the hostility between them; if, when mutual hostility is active, one group is politically dominant, it is likely to express that hostility by discriminating in economic matters against the minority; or again, if the central government or some international agency reaches down to the area where a local conflict is occurring and gives aid to the minority, the majority may become more resentful and hostile to the minority than before.

Altogether there are plenty of grounds for saying that economic discontent is a powerful force in the cumulative processes by which ethnic, religious and nationalistic conflicts develop. It may also be an ingredient in the development of hostility between nations. For where there is unemployment, deprivation and frustration, the victims may express their anger by blaming other races and nations. Hitler played upon that sentiment among the unemployed and impoverished Germans in the aftermath of the First World War, as well as upon their latent desire for revenge after defeat and for redress of the humiliating terms of the Treaty of Versailles.

If it is right that feelings of economic resentment are a common element in internal conflict and may also be expressed in hostility towards other nations, it would follow that the avoidance of mass unemployment and gross inequalities might reduce the frequency or intensity of conflict and war. As an unrepentant liberal egalitarian, that is a conclusion I like. As a social scientist, I am bound to say that, like any prediction we make about phenomena of this kind, it lacks scientific validity of the kind that can be achieved in the physical sciences. It is a judgement, no more. My basis for backing it is simply this: if I ask myself the question, would an increase or a reduction in unemployment and inequality be more likely to lead to a reduction in conflict? My answer, when I think of what I know about the history of modern conflict and what I see happening around us, is a reduction. Some people might reasonably say they do not know. Few, I think, would say an increase.

15. Lack of Parliamentary Accountability and its Effect on Arms Build-Up

Scilla Elworthy

It will be clear from other contributions to this volume how powerful, during these closing years of the twentieth century, are the factors sustaining demand for, manufacture of and trade in armaments. This chapter will examine the effects of secrecy and lack of accountability on the whole process of decision-making on defence and weapons procurement.

Decisions on major weapons systems in Britain are effectively shaped by specialists from the Ministry of Defence and from the three services, influenced by scientists and defence contractors, and are formally made by three or four cabinet ministers (McLean 1986; Miall 1987). The secrecy and accountability issues assume particular importance in post-cold war defence politics, for several reasons. First, levels of secrecy that might have been justifiable during the cold war must be re-examined now that there is no identifiable enemy. Second, as Europe moves towards new multinational modes of decision-making on foreign and defence policy, the questions of democratic accountability and control must be addressed.

British Secrecy and Accountability in Defence Decision-Making

British secrecy on defence matters is well known. Some officials explain it by saying that after the Burgess-Maclean spy scandals, the British were so anxious to recoup their good name in American eyes (in order to be given access to more US information) that heavy classification, extensive security vetting and denial of information to parliamentarians and the media became obsessive. Nevertheless, it must be pointed out that from the beginning the atomic project in Britain was shrouded in secrecy (Gowing 1988: 475).

> Mr Attlee and his closest colleagues wanted the public to know as little as possible lest an enemy learn even more. The minimum of information was given to Parliament and to the press, who for the most part accepted the situation with great, indeed excessive, docility … In practice, the estimates for expenditure on atomic energy, as on other defence research and development expenditure, were concealed from Parliament by burying them in general sub-headings of the Ministry of Supply vote — a postwar development introduced for security reasons.
>
> (Select Committee on Estimates, Fourth Report (1945/6) on The Form of the Estimates)

Later it was revealed that the funds for nuclear weapons were concealed in the Civil Contingencies Fund under the sub-heading 'Public Buildings in Great Britain'. Churchill expressed his astonishment that a statement of policy had not been made to Parliament, especially in view of the large sums of money which had been voted by the House without its knowledge (Pierre 1972: 84).

The same levels of secrecy are criticized today. Oliver Ramsbotham undertook an examination of government and opposition parties' attitudes to secrecy in 1991 and concluded as follows: 'It seems fair to say that defence decision-making is more secretive in Britain than in most other countries, insulated within Whitehall and accessible only to a handful of senior ministers, military commanders and civil servants' (Ramsbotham 1991).

This conclusion is confirmed by those actually involved in the process. Richard Mottram, as Deputy Under-Secretary of State (Policy) at the Ministry of Defence, was centrally involved in decision-making on Options for Change, about which he gave a lecture to the Royal United Services Institute (Mottram 1991):

> Essentially, the exercise was run at three levels: at the top, close ministerial control by the Secretary of State and his colleagues; a central group comprising all the top management, including single-service Chiefs of Staff, which looked at emerging proposals on their way to ministers; and a small working group of central policy and programme staffs. This approach was designed to

> ensure close central direction and a defence-wide approach, and to allow for radical options to be addressed on their merits in private, rather than in the newspapers.

Whitehall attitudes to decision-making could hardly be more clearly put. But since this exercise, the prime minister has ordered what is said to be an 'all-out attack on official secrecy and Whitehall bureaucracy'. A list of previously secret cabinet committees has already appeared, but it omits the unacknowledged miscellaneous sub-committees on defence and nuclear issues. Major has acknowledged the existence of MI6, the secret intelligence service, named its chief, and promises a White Paper to put the secret intelligence service on a statutory footing. But, in the area of defence, further than this the government will not go. The remaining areas are entirely secret:

- budgets for all weapons procurement broken down into research, development and production stages;
- budgets assigned to Top Level Budget (TLB) holders in the MoD; agreements with other nations on defence and nuclear issues, especially on research and development;
- the names of individuals responsible for specific programmes; the names of individuals responsible for errors in budgeting or project management;
- the texts of nuclear agreements with the United States and France; numbers, targetting and control of British nuclear weapons;
- the budget for, and state of development of, the tactical air-to-surface missile project;
- a definition of what in fact may be kept secret on defence and nuclear issues.

Ministerial Accountability

In theory, the checks and balances of government in Britain, without a written constitution, work along the following lines. Parliament plays a limited role in the initiation of legislation, seeing as its central duty the oversight and scrutiny of government bills developed by specialist departments for whom individual ministers are responsible. Under the British parliamentary tradition, cabinet ministers are accountable to the House of Commons. For this doctrine of ministerial responsibility to work, ministers should provide the facts necessary for members of parliament to debate the issues in an informed manner. 'The usual means by which MPs hold the Executive to account in the House of Commons is via debate and Standing Committee consideration of the government's proposed legislation. However, in the fields of defence and foreign policy, legislation is rarely involved and so MPs usually have to rely on other means' (Defence Information Group (1983–7) in Pullinger and Higgins 1989).

These other means include the following:

The Defence Estimates
These are made public in a White Paper, usually in April each year, and voted upon some six weeks later. However, nowhere do MPs see 'line items' that describe the cost of each weapons system separately in its research, development, testing and manufacture stages. Only in that way would they be able to recognize the development of new weapons systems and ask questions about them.

Parliamentary Select Committees
The House of Commons tries to monitor government actions and to obtain information through the select committees, of which two are of interest here:

The *Defence Committee* is one of 14 select committees appointed to shadow individual government departments. While it may call for evidence from and question ministers and civil servants, and recommend changes in policy and expenditure, the committee does not have the power of appropriation of funds to give effect to these recommendations. The committee regularly publishes reports of its hearings, but a substantial proportion of these are in camera and the reports are dotted with deletions of what is considered to be sensitive evidence. Ministers and civil servants questioned by the committee frequently refuse to answer questions on grounds of risk to national security. There are no publicly known or accepted criteria for the use of this justification, and information refused to members of the British parliament is often freely available in unclassified form in Washington.

In a memorandum to the House of Commons Procedure Committee in February 1990, Michael Mates, Conservative chairman of the Defence Committee, wrote that:

> There is an element of the 'Ministry of Defence culture' which appears to regard Parliament with suspicion and it is from this phenomenon that most of our difficulties have sprung. It is

> principally demonstrated in an unwillingness to answer frankly, and in what seems to be a feeling that the less that is said to the Committee, the safer the Ministry of Defence will be. On occasion this has gone beyond acceptable bounds, with the giving of evidence that concealed a serious state of affairs.[1] More frequently, it is a matter of wearisome generality or vagueness, or of trivial annoyances that make an inquiry more difficult, but which may be collectively infuriating.
>
> (Defence Committee 1990: 18).

The *Public Accounts Committee* is the most powerful committee in the House of Commons, consisting of 15 backbench MPs and chaired by a former cabinet minister from the opposition. It is the only select committee to have a large specialist staff working directly for it. The comptroller and auditor-general and his staff of 600 are independent auditors who examine the accounts of each government department to ensure that expenditure is in accordance with the law and parliament's intentions. The comptroller, however, failed to inform the Public Accounts Committee of the escalating costs of the Chevaline project, the new 'front end' for Polaris, although he had access to the relevant papers at the time. The committee was reduced to publishing an indignant report after the event.

Some more information is now supplied to the select committees on the detail of defence expenditure, but it is always supplied after the defence estimates vote, is sometimes discussed only months later, and often only in closed session. Therefore the committees are still left with little more than the opportunity of rebuke.

Answering Parliamentary Questions

One of the methods by which MPs are theoretically entitled to receive information is by means of written or oral questions. However, the answers provided to questions on defence are often uninformative, sometimes to a degree reminiscent of a Kafkaesque situation. In some instances the full provision of information might be detrimental to national security. It is, however, unlikely that these would include information on the costs and development of weapons systems. Pentagon officials are clearly comfortable with this, and are a good deal more frank than their British counterparts.

In spite of a number of parliamentary questions during the 1970s, the vital new Polaris modernization was not actually described to parliament or even mentioned by name until January 1980, 13 years after the project began, and two years away from deployment. Dr John Gilbert, Labour Minister of State from 1976 to 1979, was instructed not to refer to this Chevaline programme by name at all.

MoD officials now insist privately that the lessons of Chevaline have been learned, and that the errors will not be repeated. However, to date, a similar refusal to provide information surrounds the development of Britain's newest nuclear weapon, the tactical air-to-surface missile. During the period July 1987 to November 1989, 97 parliamentary questions on the subject were asked in the House of Commons, of which 87 received entirely uninformative replies, on the grounds that no decision had been made. Nevertheless, decisions in Britain *were* being taken; for example, the stand-off missile was the choice to replace the free-fall bomb; the new stand-off missile would carry a new British-designed nuclear warhead. Outline financial provision had also been made.[2] A Memorandum of Understanding with the United States had been signed, in the first half of 1989, sanctioning US contractors to help the UK develop the missile. This was reported in June 1989,[3] but still the House was told nothing. In fact, it is now clear that as long ago as December 1987 the British defence minister had discussed this specific programme with the Americans. In the words of Ronald Lehman, US assistant secretary of defense, in May 1988:[4]

> When your defence minister, George Younger, raised this in December, he asked us to look at a co-operative US/British programme. We thought it was a good idea. He even asked us to talk to the French and we did. And we have had good discussions. We have technical groups that have been meeting. This has been very helpful.

Although the programme has now been fully funded and the British MoD has jointly agreed technical specifications for the development of the weapon with the French MoD,[5] the British parliament in September 1993 had still been given no information.

From these and many other examples one is led to assume that ministerial refusal to provide information in response to parliamentary questions has less to do with genuine worries over national security and more with convenience, or the wish to avoid a public debate.

Defence Debates

In theory the opposition can always call for a debate on government policy. In practice, however, there are long periods when there are no debates over nuclear weapons or nuclear weapons policy at all, as for example between 1967 and 1980, a period during which plans were laid for massively increasing Britain's nuclear capability.

If the government majority were small, and if the opposition could win over government backbenchers, it could pass a vote of 'no confidence'. If that happened the government would normally be forced to resign. In practice this has not happened on nuclear issues, for three reasons. First, as we have seen, insufficient information is made available to enable a fully fledged debate on crucial decisions. In a survey conducted in 1988, British MPs from both Conservative and Labour parties said they found it more difficult to obtain reliable and comprehensive information on defence than on any other subject (Institute for European Defence and Strategic Studies 1988). In the first eight years of the 1980s, the number of Open Government Documents on defence shrank from 44 to four. Secondly, the Labour Party in opposition has been deeply split on the issue. Thirdly the Conservative Party in office has had a substantial majority, and in opposition would have been uninterested in calling a debate on nuclear weapons.

Accountability of Civil Servants

We have examined the question of ministerial responsibility. Now we must examine whether others who effectively shape defence decisions are accountable. According to Sir Michael Howard (1991), professor of international relations at Yale University:

> The characteristic which Britain shares with France, Italy and other west European democracies, is the existence of a powerful bureaucracy, the expertise and permanence of whose members make them virtually an independent element in the constitution. The extent of their influence was not unfairly caricatured in the popular television series 'Yes Minister', and is perhaps at its strongest in the fields of defence and foreign policy. Successive Labour governments, however strong their parliamentary majorities, have found it virtually impossible to make headway against the talents and institutional wisdom of this group, and have ended up diverging very little, if at all, from the course pursued by their predecessors.

In the view of Sir Robin Butler, cabinet secretary and head of the Home

Civil Service, civil servants are responsible only to the ministers at the head of their department, and not to parliament or to parliamentarians. Ministers can forbid them to appear before a select committee and when they do appear before these committees, they are under instruction to observe a code of conduct laid down by the Civil Service Department, which regulates what — and what not — to say. These regulations extend to more than 60 paragraphs of restrictions.[6] Mr Tom King correctly acknowledged, in his opening statement to the Procedure Committee, when he was examined over select committee complaints that the MoD was not providing information, that parliament does not accept these rules. The Institute for Public Policy Research recommends that they should be replaced by guidelines 'emphasizing that civil servants are bound to provide information except within narrowly defined areas of sensitivity' (International Security Information Services 1991).

As we have seen from the previous section, ministers are not, in the case of defence decisions, properly accountable to the House of Commons. The usual mechanisms to provide for accountability, namely publication of spending estimates, examination by select committees, parliamentary questions and debates are not, for the reasons given, effective when it comes to defence. If civil servants are responsible to ministers, and their duty to parliament is through ministers, and if ministers are not accountable to the House of Commons, then civil servants and others shaping decisions on major weapons systems are not accountable either.

Arguments for and Against Control and Accountability

The thorniest problem is that of access to information. Not only governments but large defence bureaucracies are reluctant to release details of defence projects and costs. They argue that such information could be of material assistance to adversaries. Opponents of parliamentary control of defence decision-making say that very few legislators equip themselves with the expertise necessary to check the executive, and that defence issues are now so highly technical as to be beyond the grasp of the average member of parliament. They say that political resolve and effectiveness, essential not only in crisis, but in piloting through large defence projects, are hampered and delayed by lengthy parliamentary deliberation. This is particularly important when it is a case of coordinating the defence efforts of 12 (or more) EC countries. Notable historians make the case that, although legislative pressures may be a brake on war-making, they may be a brake on peace-making as well, citing the British government's attempts in the 1970s to defuse the problem of the Falkland/Malvinas Islands by an agreement with Argentina, which was summarily rejected by the House of Commons, and, as another example, congressional pressures which from 1947 to 1967 prevented any *rapprochement* between the USA and the People's Republic of China (Howard 1991).

Those who advocate parliamentary accountability have seven main points to make:

(1) In the absence of informed criticism, a small group of decision-makers could be making bad decisions, but since no 'outsiders' would know, no one could point out deficiencies or errors.
Harold Macmillan warned 20 years ago that cabinet ministers are rarely equipped with the technical expertise to make an adequate assessment of the validity of a new weapons system (Macmillan 1972). The technology to which he was referring has grown yet more complex in 20 years; and with the sheer pressure of work involved, ministers tend to be more concerned with the presentation of policy and the deflection of criticism than with actually making the decisions. They tend to accept the predigested options which are presented to them by a small and isolated group of experts who may have been working on a particular system for a decade or more. Their decisions, because they are secret, do not receive the criticism of their peers, traditionally regarded as a vital element in scientific endeavour.

Lord Zuckerman, chief scientific adviser to the MoD and the government for more than a decade, said, 'Discussion of highly important technical matters with strategic and economic overtones within a small circle that not only has a vested interest in the subject, but which is also shielded politically from informed and critical debate, does not necessarily result in advice that leads to the wisest political decisions.'

(2) Money is wasted.
The most obvious example is that of Chevaline, but there are countless others. The initial cost estimate for the Chevaline programme was £250 million and the project ended up costing more than four times that. The problems involved not only managerial inadequacy on the part of the MoD, but also political indecision on the part of those Labour and Conservative ministers who knew about the project.[7]

(3) Alternatives are not investigated.

If proper information is not provided to the House, parliamentarians are rendered incapable of debating defence policy. As a consequence there is no department within the MoD which carries out research into alternatives to existing policy; for example, there is no in-depth examination of how a non-nuclear defence policy for Britain could be carried out and no costing of such a policy.

(4) The result of not investigating alternatives (see 3) and restricting decisions to very few and largely non-elected personnel (see 1) is the tendency to 'carry on doing the same'.

In this way, decisions on individual weapons systems become decisions on policy; policy is predetermined by incremental decisions on new weapons systems each of which, given the way that decisions are made, appears totally justified.

(5) Lack of informed public debate.

The restriction of decision-making to very few and largely non-elected personnel (see 1) and the refusal of information to parliament, means that informed public challenge to decisions is rendered impossible and that decisions are made which are either incomprehensible to or unsupported by the electorate. The diagram on the following page is a graphical illustration of this point (Oxford Research Group 1991).

(6) Vested interests.

Real accountability is the only way of counteracting the influence of vested interests. Obvious and powerful vested interests in the continuation of the arms race expend large sums of money in lobbying decision-makers to decide in favour of a new weapons system or a continuation of present policy. Many of these vested interests, as we have seen, are privy to information not available to members of parliament. There are, however, no vested interests in stopping the arms race, or in stopping a particular weapons system from being developed. The sums of money available to organizations like the Campaign for Nuclear Disarmament, the Campaign Against the Arms Trade, and Scientists Against Nuclear Arms all have to be raised from public subscription, and are a minute fraction of the sums spent by defence corporations and government agencies such as the Defence Export Sales Organization.

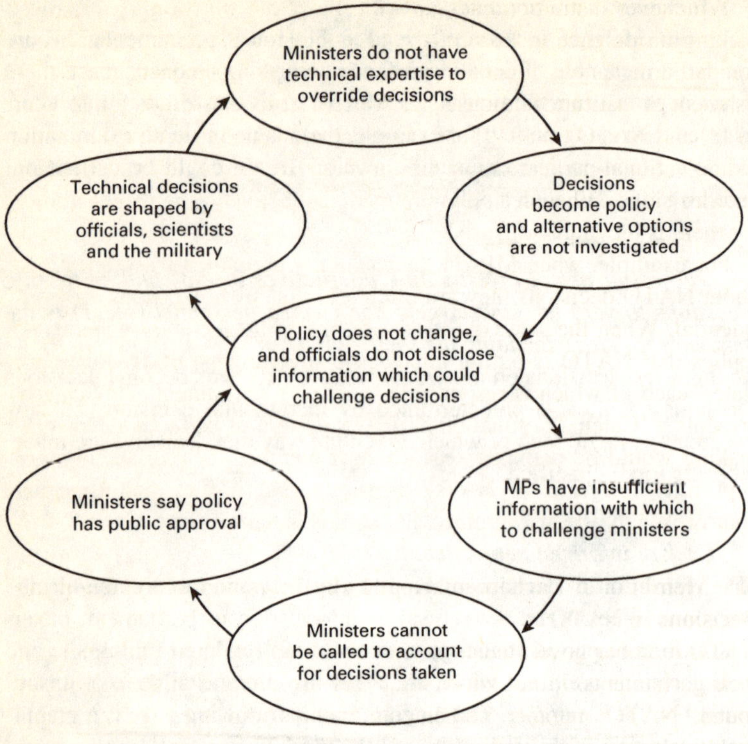

(7) Decisions satisfy technical criteria.

Traditionally, defence decisions tend to be influenced most by technicians and military experts, due to very long lead times, and the relatively short tenure of politicians. This means that defence policies are oriented in the direction of weapons and kill-ratios, rather than towards political factors and war-prevention techniques.

Multinational Defence and Foreign Policy Decision-Making

Fundamental to the transformation of Europe over the past two years and expressly placed at the centre of the agenda in both NATO's London Declaration (July 1990) and the Conference on Security and Co-operation in Europe (CSCE) Paris Charter (November 1990), is the principle of democratic accountability. 'The participating States declare that the will of the people, freely and fairly expressed through periodic and genuine elections, is the basis of the authority and legitimacy of all governments.'[8]

Whichever institution ends up with the prime function of decision-making on defence in Europe, the necessity for adequate mechanisms for parliamentary accountability becomes pressing. Indeed, the establishment of institutions such as the Western European Union (WEU) (in 1948) and NATO (in 1949) has gradually weakened the control by individual national parliaments of their own governments' defence budgets and decisions. When decisions are made collectively, the responsibility for them is inevitably diffused.

For example, when MPs in their national parliaments ask questions about NATO decisions they are often told that such decisions are confidential. When the same questions are put to the secretary-general, he replies that NATO is but an alliance of governments of 16 sovereign states, each of which is responsible to its own parliament. This clearly presents a 'Catch 22' situation. NATO's system of collective decision-making would be perfectly reasonable if members of parliament were kept fully informed of NATO decisions, and if they had financial control. We may now examine whether this is the case.

Are Members of Parliament Kept Fully Informed of Forthcoming Decisions in NATO?

NATO member governments are not answerable to their parliaments if their parliaments do not know what they are doing. Put the other way round, NATO member parliaments cannot hold their governments accountable to them if they are not given full information at a time when it is still possible to affect the outcome of decisions.

Of course, there are within NATO grounds for the withholding of information for security reasons. Each member state of NATO has subscribed to an agreement binding the parties to ensure that NATO classified information is protected as appropriate and, in particular, is not disclosed to another nation without the consent of the originator. In NATO's view this is imperative, because otherwise all mutual trust between allies would be forfeited, frank and open discussion would become impossible, and the alliance would not be able to present the united front essential to its policy of deterrence.

Having said this, it should be the case that all allied parliaments can receive the same information. This is not the case. Take, for example, the US plans to deploy in Europe new nuclear artillery shells of the neutron bomb type. British defence ministers denied in parliament in late 1984 and early 1985 that any such proposals for 'modernization' had ever been made, let alone accepted.[9] However, seven months earlier, Dr Richard Wagner, assistant to Defense Secretary Caspar Weinberger, told a congressional committee that as long ago as October 1983 the new 'neutron-capable' 155 millimetre and 8 inch nuclear artillery shells had been endorsed by NATO ministers.[10] He was asked in April 1985 by a congressional committee, 'Is their [i.e. the British ministers'] desire not to be explicit *for home consumption*?' He replied, 'Probably.'

Attached to NATO and WEU are the North Atlantic Assembly and the WEU Assembly. Each consists of MPs drawn from member-state national parliaments, providing a forum for international discussion of matters relating to NATO and WEU. They have no formal influence or control over the decision-making in their institutions. However, because they are independent to some extent, such assemblies can set an example. After the 1989 European revolutions, the North Atlantic Assembly was quick to invite representatives from Eastern Europe.

The European Community, the Commission, the Parliament and the Council

The Maastricht Agreement of December 1991 extends the powers of the European Union ultimately to include decisions on defence and foreign policy. Many members of both the European Parliament and national parliaments, however, doubt the powers of the European Parliament to monitor, let alone control, such decisions. If the EC is to undertake defence decisions where eventually military forces are involved, the parliament must have the powers to exert democratic control over such decisions, including budgetary control.

Do Members of Parliament have Ultimate Financial Control over Defence Decisions?

The ultimate sanction of parliament is to refuse to grant money for projects it does not accept. In order to do this, a parliament needs to know how much money its government is committing to a particular project, and when.

Only one NATO parliament, the US Congress, has a defence budget which is itemized in such a way that representatives can see how much money is being spent on a particular weapons system. This means that they have some idea when a decision has actually been taken, because they have to vote the money for it.[11]

Turning to the question of national contributions to NATO's own budgets, and accountability for these, it proved impossible for the author to obtain any figures from NATO itself. Over a period of four months from April to August 1992 repeated requests were made to NATO information offices to provide details of military and/or civil budgets. All NATO offered was, in its own terms, 'a broad generalization', namely that the overall total for civil and military budgets is approximately $2 billion annually. Of this, the infrastructure budget is approximately $1 billion, and the other three budgets are the civil budget, the budget for the working of the Military Commission (these last two are those referred to on p. 222) and lastly the budget for the AWAC's squadron based in Luxembourg. NATO refused to supply any further information, saying that it was classified and going to the extent of denying the existence of figures on national contributions in any collated form.

Only extensive research in the member states permitted approximations of the figures to be obtained. This, and the fact that these figures are neither published nor available in the House of Commons Library, leads to the conclusion that they are not known to British members of parliament. Initial enquiries in other NATO countries indicate that other parliamentarians may be similarly unable to obtain information. Such difficulty in obtaining information presupposes the impossibility of control over decision-making.

Thus, the national parliaments, except Congress, lack adequate powers properly to control defence expenditure. Moreover, the multinational parliamentary assemblies of defence organizations, namely the WEU Assembly and the North Atlantic Assembly have no powers of budgetary control at all.

Conclusions and Recommendations

Decision-making on defence today is a very different challenge even from that of a few decades ago. Strong democratic control of decisions, with full parliamentary debate and accountability, becomes important for four principal reasons. First, new weapons now tend to dictate new military strategies; parliamentarians must be aware of new weapons in the pipeline before it is too late to cancel. Second, the complexities and the costs of equipment are soaring, and mistakes can cost millions. More informed and efficient oversight becomes vital. Third, decision-making on weapons procurement is increasingly a multinational affair; accountability structures and control mechanisms should follow. Fourth, the geopolitical map of Europe has changed out of all recognition over a short period. New demands and priorities are emerging daily, which cannot be met by 40-year-old military principles and security concepts. The response has to be dictated by political rather than by military considerations, and controlled by democratically elected representatives.

Ministers and senior politicians from countries of Eastern and Western Europe met in May 1991 in the Netherlands to discuss these issues, at the initiative of the Netherlands minister for foreign affairs, Hans van den Broek. In his summing-up, the chairman of the conference noted some practical suggestions made during the conference for improvement of parliamentary control of decision-making. These included:

- Defence budgets to include line items showing the estimated expenditure on the stages of research, testing, development, prototype production and full production for each weapons system. Only in this way can parliamentarians become aware of, and approve or reject, new weapons developments.
- A definition to be arrived at, with the aid of a special committee of elected members, of precise categories of information which may justifiably be kept secret. Legislation may then be necessary to enable members of parliament or citizens to obtain information unnecessarily kept from the public domain (which might be modelled on the US Freedom of Information Act).
- Agreements with other nations, including those for collaborative procurement projects, should be subject to full parliamentary debate before signature.
- Specialist committees should have the power to compel officials and military staff to appear before them for questioning. The information which they are entitled to receive should be governed by the second item above, without which parliamentarians cannot have an informed view on which to base their votes.

Many noted studies (for example, Marsh and Fraser 1989) have shown that public opinion in Britain generally deplores waste on unnecessary weapons, and supports a lower level of armament than that which exists, as well as lower arms exports. Members of parliament say that they cannot get the information they need in order to challenge new arms projects or levels of export. The whole arms export scandal surrounding the Iraqi supergun project, and its subsequent investigation in the Scott enquiry, illustrate only too literally how parliament and the public can be and are kept in the dark over what Britain exports and to whom. The United Nations 1993 arms register is certainly a step in the direction of greater openness, but in the absence of binding international agreements, there is nothing at present to hinder unrestrained worldwide competition in arms. As far as British national decisions are concerned, until parliamentary procedure is changed to enable members of parliament to obtain the information they need to act on constituents' wishes, the present situation is likely to continue.

Notes

1. The 'concealment of a serious state of affairs' referred to the Trident Nuclear Submarine Programme.
2. On 24 May 1988 George Younger told the House of Commons 'We have allowed a generalized amount of money in our forward costings for replacement, but we cannot do anything accurately until we know broadly what the cost of the systems will be.'
3. *Jane's Defence Weekly*, 24 June 1989.
4. Interviewed on BBC *Panorama*, 3 April 1988.
5. Information from French Ministry of Defence official, 6 June 1992.
6. Cabinet Office, March 1988.
7. The examples are well documented in the findings of the public accounts committees of parliaments all over Europe and North America. In the case of the UK, critics cite also the Nimrod Early Warning System (conversion of 11 anti-submarine planes to radar-capability: development costs escalated from £34 million to £980 million before the project was cancelled in 1986); the Type 23 Frigate Computer (eventually cancelled in 1987 after expenditure of £30 million), and the waste of £18 million introducing computer systems for the MoD which failed to deliver.
8. Supplementary document to give effect to certain provisions contained in the *Charter of Paris for a New Europe*, 16 November 1990.
9. Parliamentary questions, 24 October 1984, 29 October 1984, 28 January 1985, 26 February 1985.
10. Congressional testimony, 1 May 1984.
11. In practice, however, the Pentagon has a substantial amount of money which is used to research or continue specific projects before Congress has actually voted money for them. This is known in defence circles in the US as 'the Black Budget', and was estimated by informed commentators to amount to $24.8 billion dollars in 1989 (Centre for Defence Information, Press Release, 21 March 1988). Secret spending of this kind paid for the development of weapons systems such as the 'Stealth' Advanced Technology Bomber, the Advanced Cruise Missile, and the Advanced Tactical Aircraft. 'There are instances where secrecy is necessary,' comments Rear Admiral Eugene J. Carroll, US Navy (retd), 'but far too many programmes are unnecessarily cloaked in secrecy. Major weapons programmes must be subject to informed public and congressional scrutiny if we are to ensure that we are buying only the weapons we need to provide an effective defence and that we are paying a fair price for those we buy' (Centre for Defence Information, Press Release, 21 March 1988).

16. The Arms Trade

Michael Brzoska

The international trade in arms is a major feature of international politics and economics. In the early 1990s, it had a volume of about US$ 30 billion annually, down from a peak of almost US$ 50 billion in the mid-1980s. More than half the arms trade is exported from the north to the south. Its share in the total import of developing countries is about 4 per cent. About 90 per cent of all weapons in the arsenals of developing countries were produced in the industrialized countries of the north.[1]

Given that poverty is widespread in many of the recipient countries, the financial costs of importing arms, over and above the immediate costs in terms of foreign exchange, are large. Arms are often delivered on credit terms and, without arms imports, Third World debt could be lowered by more than 20 per cent. In the 1990s so far, more money has been spent on repaying debts for military imports than on buying new weaponry (Brzoska 1994).

The costs of arms imports can go beyond financial issues. In times of war, arms are used to bring about devastation, to kill people and to damage the environment. Imported weapons used in wars are a visible link between war and the arms trade, which is obvious to every television viewer. Beyond simple observation, there is a widespread belief that weapons imports are a major factor in leading to wars. This belief was already strong among liberal intellectuals before the First World War; it became dominant, even within the governments of many countries, after that war.

But the arms trade not only continued; it grew almost yearly after the Second World War until it reached its peak in the mid-1980s. What makes the arms trade such a persistent feature of international relations? Who are the actors and what are the forces sustaining this trade?

It is quite useful to think of the arms trade as an institution in a twofold sense. It is an institution in the sense that it represents generally accepted behaviour in the international society of states. It is also an institution in the sense that specific roles are held by specific actors. The producers of weaponry, the armed forces, the arms dealers, the governments of exporting and importing countries are all indispensable in keeping the arms trade alive. At the same time these actors feel restricted in their freedom to act because a specific arms transfer system is shaping their behaviour. Or, put another way, structures and interests combine to support the arms trade.

I will look first at the actors, or rather stylized types of actors, and then at the arms trade as an aspect of the international system of states. Of course, institutions derive their strengths from the mutual reinforcement of roles and rules, so the two levels cannot be strictly separated. I will go on to consider specific links between the arms trade and wars, since concern about the latter is what this book is about. In conclusion, I will briefly review approaches intended to weaken the arms trade as an institution of war.

A short contribution such as this cannot deal with the full complexity of the subject. Issues such as the proliferation of weapons of mass destruction, or the transfer of dual-use items, an increasingly important element in the transfer of military related goods, are not treated. The focus is on the trade in conventional weapons with recipients in developing countries.

Actors in the International Arms Trade

Arms Dealers

The classic culprit in the arms trade is the arms dealer. The prototype of all arms dealers is Basil Zaharoff, who sold guns and ammunition for Armstrong-Vickers before the First World War. His success in acquiring customers worldwide, sometimes on both sides in a regional arms race, contributed much to the early popular sentiment against the arms trade. The image of the shady dealer without morals, willing to sell to everybody and his enemy, can still be easily recreated in the public mind.

But there is a certain amount of ambiguity in this respect. Clearly, private arms dealers are a part of the arms business. The Iran–Iraq war of 1980–88 was their bonanza. But, as for instance Sam Cummings, probably the best-known of today's arms merchants, insists in his interviews, if one wants to stay in the business one has to work within the frameworks established by governments; otherwise an arms dealer is soon in gaol. Private arms dealers often work in the shadow of governments. A good example was the Iran-Contra affair in the United States. Every one of the various arms dealers from Israel, the USA and Iran acted under the protection of at least one government. The use of private dealers gives governments the option of a 'plausible denial' if things go wrong.

Today, arms dealers are mostly found in the grey areas of the arms market. The sizes and shades of grey of these markets are dependent on the extent of governmental undercover activity, and of the divergence of official government policies from actual policies.

Arms Producers

Today's arms sellers are predominantly managers in arms production companies, respectable gentlemen who despise the image of Basil Zaharoff and the like. For them, selling arms is a logical extension of their business of producing arms.

Arms producers have great commercial interests in arms exports, interests that go beyond the immediate income from foreign sales. Arms exports increase the number of items produced. Since production runs in the arms business are often short, costs per item decrease fast with more customers — a saving that can be either pocketed by the producer or used to increase competitiveness. For instance, in 1991 the British government awarded the contract for up to 300 new battle tanks to Vickers only after Vickers quoted a price for their Challenger II tank based on the expectation of extensive sales. Without exports (or rather expected exports in this case) Vickers would have lost the order and would most probably by now be out of the tank-producing business altogether.

Arms exports are also an important argument in the sales strategies of arms producers. Producers promote their products at numerous weapon exhibitions all over the world and in the thriving arms trade media industry, which is financed by the arms producers' advertisements.

Armed Forces

Leaders of armed forces and their civilian couterparts in defence ministries tend to have an ambiguous view of arms exports. On the one hand, exports have the potential to lower domestic procurement costs. They also have some symbolic value, as they demonstrate the quality of weapons that are stocked in the national arsenal. On a more practical level, weapon use in conflicts provides useful insights into their capabilities and shortcomings.

But arms carry military capabilities, which (and not only from the point of view of generals) should not be in the hands of actual or potential enemies. Armed forces therefore can be expected to plead against the sale of the most modern weapons, especially if the chances are high that transferred weapons will be used against forces from the producer country. That happened, for example, to British soldiers in the war over the Falkland Islands in 1982 and to French and US troops during the second Gulf War of 1990/1. In the post-cold war world, with an increased number of foreign military deployments, such encounters may become more frequent.

Exporting Governments

The state is nowadays central to arms exports. In fact arms-producing states have imposed laws and regulations on the export of arms ever since the 1920s. Today almost all states have regulations concerning the export of arms. States also regulate production, both through laws requiring licences to produce arms and because arms producers are usually financially dependent on their national governments.

Governments have strong interests in maintaining strict control over arms exports. These interests can be discussed under the headings of restriction, power and economy.

Restriction is the most ambiguous of the national interests in arms exports. If taken literally, some of the laws on arms exports seem to have restriction as their major objective. For instance, the still valid French *Décret-loi du 18 avril 1939, fixant le régime des matériels de guerre, armes et munitions* reads in article 12 that all exports of arms are prohibited except in the case of exemption authorized by the authorities. Such sentences should not be misinterpreted. The purpose here is not restriction, but control. The French government, for instance, is granting export licences rather freely, giving France a large share in the international arms market.

Still, some types of actual restraint are in the interest of exporting states. First, nobody wants to make enemies stronger, so exports to actual or potential enemies are restricted. During the cold war, the West tried to achieve this through the Coordinating Committee on East–West Trade (COCOM). COCOM did not disappear with the cold war; export controls are now redirected towards the south. Second, restrictions in arms transfers are sometimes used as signals to other states. For instance, Britain claims not to export arms to states with gross violations of human rights.

Some states have more far-reaching restrictions. A number of neutral European states, such as Sweden, Austria and Switzerland, but also Germany, prohibit the sale of arms to countries at war. The state with the strictest export rules is Japan, where no weapon exports are allowed.

The power motive has many facets. Arms transfers can be used to achieve specific objectives, such as base or overflight rights or the release of hostages. Arms transfers create and cement alliance relationships. Arms transfers are a subsitute for direct military involvement. Arms transfers are an instrument for demonstrating support for specific policies in a recipient country.

On a more abstract level, arms transfers can be linked to the 'three faces of power' distinguished by Steven Lukes (1974). First, arms are used as a means of bargaining power to achieve wanted outcomes directly. Other states are paid to do what an arms exporter wants in the currency of weapons. Second, arms are instruments of structural power limiting the manoeuvring space of recipients. Arms deliveries create and cement dependency relations, since it is costly to switch from one supplier to the next. Finally, arms are related to situational power. Weapon systems are more than compositions of metal and electronics. They are expressions of a type of technology, of a way to organize cooperation and hierarchy in an organization. They are the epitome of the military in industrialized countries and when transferred proliferate that model (Kaldor 1982).

Arms transfers are also an instrument of economic policy. Again, there are many facets. Arms exports improve the balance of payments. They provide employment in the cities and regions, often in economically difficult areas. Arms are mostly sold by producers in high-technology industries so that exports increase the funds available for domestic research and development, and thus further competitiveness of domestic industries. Arms exports can make procurement for the national armed forces cheaper. In some cases, such as the French, full-scale domestic production would not be viable without arms exports.

So governments have strong interests in having good control over arms exports, but interests in restraint vary. Factors explaining actual arms export policies of governments include the state's position in international politics (big powers are dominating international politics), economic dependence on arms production, historical experiences (as in the Japanese case) and domestic moods.

Importing Governments

It is a widely accepted assumption that the demand for weaponry is unlimited, or, rather, limited only by the economic means available. This, of course, is a rather strong limitation for many importers of weapons in developing countries. In fact, differentiation of interests and actors within importing states leads to a somewhat different conclusion.

In poor countries, armed forces can be expected to argue for unlimited importation of the best weapons available. The more modern the weapons available, the higher the probability that they can get their military tasks done. High-tech weapons are preferred by general staffs and procurement officials even when their soldiers are not well prepared to cope with them. There is a general, often ill-advised, preference for modern weaponry among today's armed forces. The reasons are manifold, ranging from the perception of superiority, reinforced by wars such as the second Gulf War, to the assumption that the privileged use of high-tech gives soldiers an elevated status within their societies. Domestic industry, even where it is sizeable, is unlikely to produce weapons of the quality of the major exporters. Also, because of the shorter production runs, domestic production is generally more expensive in poor countries, unless labour costs are an important element in the price (Brzoska and Ohlson 1986).

In a way it is the job of generals to argue for more; but their demands should be subjected to scrutiny and counterbalanced with demands from other groups in society. Unfortunately, in many poor countries the generals are the government. Luckily, this does not always mean that the armed forces get whatever they want. Empirical studies have shown little correlation between the type of government and the level of arms imports (West 1992). Civilian governments are afraid of being unseated by generals and try to keep them busy and happy by importing expensive weapons; generals in power have had larger aspirations than to modernize their armed forces and kept arms imports down in order not to endanger their economic policies.

In underdeveloped economies and political structures, budget-balancing is a tricky process, but one that gives the armed forces some advantages, since they are generally the best-organized and most coherent organization. The economic argument that spending less on arms imports would be better for the economy has no comparable proponents. In fact, it is nowadays often outside development aid givers and international financial institutions, such as the World Bank and the International Monetary Fund, that raise the issue of military overspending with domestic governments (Ball 1992). The response, so one is told in private by participants in such endeavours, is mixed. Heads of state and their ministers of defence (or of war in a few cases) are angry. Finance ministers keep their mouths shut, but later thank the foreigners for making the arguments they always wanted to but were too afraid to make.

Arms Transfers in the System of States

The 'Good Importers' Dilemma'

If generals and ministers of defence are smart, they use arguments the experts from the World Bank or development-aid agencies would find difficult to counter. Sure, they would agree to stop importing arms — but only if their neighbours did the same.

If the generals and ministers had read strategic studies literature, they might argue that they were caught in a 'prisoners' dilemma' situation. On the one hand, they and their enemies would be better off if they imported no arms. But, unfortunately, defection would pay: if only one state disarmed, the other one could exploit that situation. Also, they could point out, they lived in an 'anarchy', an international system in which each state had to look after itself, where the responsibility to survive was only with the state itself, and not with some international body. Even the 'good' state that has no designs on any of its neighbours cannot avoid the prisoners' dilemma and has to do what the system of states prescribes.

... and the Reality of Arms Imports

The 'good importers' dilemma, though considered the climax of theory in the mainstream of security studies, has only limited relevance in the actual politics of the international arms trade. Direct links between the arms imports of one state and a neighbouring state are more the exception than the rule. They occur in the highly conflictual situation in the Middle East, for instance, between Israel and Syria, and in South Asia, between Pakistan and India. Even in these cases, it is of course not necessarily the case that one of the states would have perished if it had not imported arms.

Often, arms imports are very difficult to relate to the threat situation of a state. The Shah of Iran still provides the best example. He personally ordered large quantities of high-tech weaponry, such as US-American F–14 Tomcat fighters, for which there were few Iranian operators. The aircraft were symbols of superiority of a Pahlavi over the rest of the world, and, for the Shah's enemies, of waste and of dependency on the United States. When Saddam Hussein ordered an attack on Iran in August 1980 it was not the Tomcats, but 'human waves' of young, untrained soldiers that beat his troops back.

In some cases, arms imports helped create those threats that were later used to justify more arms imports. For instance, when Argentina started its large modernization programme in the mid-1970s, which cost it more than US$ 20 billion and — via indebtedness — is still a major burden on the economy, there was no neighbour threatening its territorial integrity. Instead, later on Argentina's armed forces almost fought a war with Chile in 1978 and lost a war with Britain in 1982.

The 'Good Exporters' Dilemma'

There also is a 'prisoners' dilemma' type situation on the exporters' side, linked to the international economics of arms sales.

There are good economic arguments against arms sales. For instance, it can be shown that investment in military research and development is less productive than investment in civilian research and development. The Austrian economist Kurt Rothschild (1973) demonstrated in a classic econometric study that those countries with comparatively large arms exports (USA, France, Britain) had lower growth rates in civilian exports than others (Germany, Japan). Also, military sales take away buying power from civilian products. Money that has been spent on imported weapons cannot be spent on the imports of civilian goods.

So everybody would benefit from reduced arms sales — but, as in the prisoners' dilemma, coordinated behaviour is difficult to bring about because defection pays. If nobody sold weapons, world trade would be larger, but if only some stopped the sale of weapons, those continuing arms sales would increase their market shares and those stopping sales would not gain civilian exports.

... and the Reality of Arms Exports

The argument of 'if we do not sell, somebody else will' is a quite powerful one. There is large overcapacity in arms production. On the international arms market recipients can choose between suppliers. Recently, they predominantly chose US suppliers.

The USA is by far the largest arms exporter in the 1990s. Though economics have become more important with the decrease in domestic procurement, the USA is still the classic hegemonic supplier. Another advantage of US suppliers in the international market derives from the size of the budget for military research and development in the USA. It is more than double that of all West European states. That translates into a technical superiority of US weapons in many areas.

Russia's producers are desperately trying to increase arms exports, since they are more or less the only industrial goods that can be sold on the world market. But recipients are reluctant to buy from the former Soviet Union, even though prices are very low. A major reason is the problem of long-term supply, another technology. But it should not be overlooked that the reduction in Russian sales is partly fictional: a large part of the exports from the Soviet Union were to allies and friends in poor countries who were not really expected ever to pay. Now, with a strictly commercial arms export policy, such deliveries no longer increase Russia's share in international arms statistics.

The arms market has long been dominated by the USA and Soviet Union. From the 1970s, more and more producers appeared, first in Europe and then elsewhere in the world. Lately, these suppliers have come under economic pressure. Producers from Western Europe have lost some of their business in the arms market. Efforts to counter the trend focus on the creation of a West European arms market under the aegis of the European Community. Producers from other countries, so called third-tier suppliers, lost even more. Most hard hit were typical low-cost suppliers, such as Brazil and Bulgaria. They are least well equipped to compete with the USA in high-tech weaponry.

Arms Transfers and Wars

The relationship between arms transfers and wars merits a somewhat more detailed discussion. Wars obviously create an important demand factor for arms sales. Also, arms transfers are sometimes a factor leading to wars and, almost always, in their continuation. In some cases, arms transfers have facilitated the ending of conflicts.

War as a Factor of Demand and Supply

Some wars have had major impacts on the level of the arms trade. The 1973 war between Israel and Arab states, for instance, although very short, was accompanied and followed by large-scale deliveries of weapons. The Iran–Iraq war was a major boon for arms exporters from all over the world. Close to 20 per cent of all arms exports went to these two countries during 1980–88.

It has been estimated that about one-third of the total delivery of weapons between the early 1960s and the late 1980s went to states at war (Brzoska and Pearson 1994). It has never been estimated how many of the weapons delivered have actually been used in war. Since most of the weapons used will have been delivered earlier, the ratio is probably much higher than one third of all the weapons delivered.

Weapons from all exporters have been used in wars, but some exporters' weapons were especially likely to be used in combat. One aspect of the arms trade, at least since the 1970s, has been a shift away from traditional suppliers in times of war. This trend is not universal. Israel, for instance, has relied on the United States as almost sole supplier since the late 1960s. But in other cases, relations between traditional suppliers and recipients were substituted by sales from new suppliers in times of war.

The most striking example was provided during the Iran–Iraq war. Originally, Iran had relied mostly on the United States for weapons deliveries, with some weapons coming from countries like Britain and Germany. Iraq's dominant supplier was the Soviet Union. During the war, military equipment from more than 50 countries was delivered to the two warring states, with more than 20 names appearing in both lists of suppliers. The Soviet Union was the only one of the earlier major suppliers to continue to play an important role, though its sales to Iraq were much less important than before the war. Instead, a large number of suppliers from Western and Eastern Europe, Latin America, Africa and Asia provided the equipment for the war, including Brazilian tanks, South African howitzers, Chinese missiles and Swedish fast-attack boats.

Weapons as a Factor in Wars and Peace Diplomacy

Obviously, the current and perceived level of armaments plays a major role in the decision to go or not to go to war. Accordingly, arms transfers can influence that decision. Since they can do so in many and diverse ways — both strengthening the resolve to go to war and creating a situation of deterrence — no general conclusions on the relationship between arms transfers and the onset of war have been empirically validated. Of course there are individual cases in which arms transfers played a definite role, such as the decision by the Argentine junta in late 1981 to go to war with Britain if negotiations over the Falklands/Malvinas did not succeed soon.

Suppliers often try to use arms transfer relationships as a means to influence war outcomes. A comparative study of a number of wars has shown that in most cases efforts by arms sellers to influence the course of wars and peace negotiations were futile. This is true for short wars, such as the 1982 Falkland Islands war, because the needs for resupply and more weapons generally remain limited. It is also true for longer wars, if warring parties have access to other sources of weaponry, as in the case of the Iran–Iraq war. The impact of outside supply on wars was greatest when resupplies were massive, as in the 1973 Arab–Israeli war, or in the 1977/8 Ogaden war between Somalia and Ethiopia. Outside weapons supplies during wars did in any case increase the level of fighting and destruction. Arms sales may not lead to more wars, but they make wars bloodier.

Arms Transfers and the Future of Warfare

The possibilities for switching suppliers increased during the 1970s and 1980s with the appearance of more suppliers. Possibilities for major suppliers to steer the outcome of wars decreased correspondingly. Although some of the smaller suppliers have disappeared in the 1990s, we are not back to the situation before the 1970s, but actually in a quite new situation.

There have been such large supplies of weaponry that there are hardly any groups wanting to fight a war that cannot find the weapons to do so. Of course, it is easier to get mortars and rifles than artillery and tanks. More than 500 million automatic rifles have been produced in the last 50 years or so and most of these remain in service. If peace is achieved in one region, weapons are distributed to the next war. For instance, peace in Ethiopia from 1991 aggravated the situation in neighbouring Somalia and Sudan, which were flooded by weapons — many of them originally delivered during the Ogaden war of 1977/8.

For small arms, the time for limitation of weapons is probably over. There are too many around. They fuel the large number of wars that are currently occurring in the developing world. There is little chance of controlling them, but also little sense in increasing their numbers.

Limiting Arms Transfers

How can the flow of weaponry be limited? With strong interest groups in both recipient and supplying countries, what is the best approach to limiting the damage in terms of economics and destruction in war? I will briefly describe some of the approaches, sorting them by their links with major approaches to thinking about international politics.

Legalistic Approach

When after the First World War popular resentment against the arms trade was high, diplomats and international lawyers set out to abolish the institution of the arms trade. The covenant of the League of Nations, adopted in 1919, authorized the League in Article 23 to supervise the international trade in weapons. This was attempted in the St Germain Convention of 1919. The convention never came into force, mostly because the United States refused to become a member of the League and, consequentially, to sign the convention. After this failure, the League decided to make less ambitious initiatives. One concerned the gathering of statistics and information on armaments. From 1924 to 1939, the League secretariat published *Armament Yearbooks*, in which publicly available data on military expenditures, force levels and the production and trade of goods with military relevance were gathered.

No similar large-scale efforts on the legal front were made after the Second World War. The issue of the arms trade was considered to be a secondary one to the issue of war. If war could be abolished — after all the ultimate purpose of the United Nations — the arms trade would automatically cease to be dangerous. Conversely, efforts to stop the arms trade would not succeed so long as war was not abolished.

Those legal restrictions on the arms trade that were developed after the Second World War refer to special types of weaponry and equipment, such as the Inhumane Weapons Convention, the Nuclear Non-Proliferation Treaty, the Biological Warfare Convention and the Chemical Weapons Convention. The 'Friendly Relations' Declaration (UNGA 25, Res. 2625) formally forbids arms transfers to recipients other than governments, unless the non-governmental recipient is fighting colonialism, racism or gross violation of human rights. The UN Security Council can decide on arms embargoes (Art. 41 UN Charter). Before the 1990s, it did so in a few cases, most notably against South Africa. After the end of the cold war, a number of arms embargoes were declared, such as those against Yugoslavia, Liberia and Somalia because of internal wars, against Libya because of support of terrorism and against Iraq because of aggression.

One other development should be mentioned here. In 1991, the UN General Assembly decided unanimously (with two abstentions) to create a UN register of the transfer of major weapons. Almost all suppliers and a good number of recipients sent in reports about their arms deals for the first report in 1993. This register is typical of the state of international legal approaches to the arms trade. They are concentrated on the fringes of the trade, on those weapons and countries deemed especially dangerous. But in general the objective is not limitation but more transparency and better control over what is happening.

Realist Approach

The dominant post-Second World War approach is an expression of what in international relations theory is called 'realist' thought. Said differently, it is not arms that are the problem, but the conflicts underlying the demand for weapons. In the realist perspective, the only promising way to limit the arms trade is to solve the underlying conflicts, or at least to reduce their intensity. This can be done partly through negotiations between conflicting parties, but also through 'great power' intervention. For realists, it is idealistic to assume that limitation can come from the supplier side. There are too many suppliers and their interests in arms transfers, economic and political, are too powerful. So the burden is on the recipients.

Liberal Approaches

The great theoretical rival to realism, often called 'idealism' or 'liberalism', rejects the deterministic view of states' behaviour and stresses the importance of decision-making processes within states. In the liberal view the citizens and not the state are the prime actors.

Periodically, the arms trade has been discussed in many of the supplying countries. Often such debate was the result of some specific 'scandal', such as the Iran-Contra affair in the USA, or the 'supergun' case in the UK. Typically, arms transfer restrictions are tightened after such scandals — but often only for the time being, until the outrage has subsided.

Compared with the public interest in arms transfer matters, it is surprising how decision-making on arms transfers is concentrated in the realm of the executive. The United States is alone among major suppliers in having a legal role for the legislature in decisions about arms transfers. In a few countries, such as Italy and Sweden, parliament is regularly informed about arms deals after the fact. It is probable that more parliamentary involvement would result in more restriction (see Elworthy in Chapter 15).

Neo-Marxist Approaches

Is it really conflicts that are the source of the demand for arms imports? If so, what are the sources of conflicts? Is not demand stimulated by the activities of the great powers? Is it not their desire to be militarily superior that drives the armaments process in recipient countries? Are not the officers ordering these weapons graduates from military academies in the supplying states?

Authors in the 'dependency school' tradition have not only claimed that arms are 'pushed' from supplying states, they have maintained that the conflicts are mostly a product of the doings of the industrialized countries. They singled out the world economic order, which works to the detriment of most inhabitants of the south, as the main cause of conflict.

In this view, arms transfers limitations have little prospect if they are not combined with improvement of living conditions in many regions of the south through a more just world economic order. This twofold approach is said to be reinforcing, since reducing arms imports increases the chances for more economic development.

The Future of Restrictions on the Arms Trade

Rather little is happening in arms transfer control today. Prohibitions concentrate on some weapons and some states — those that are most dangerous to a world dominated by the major suppliers of weaponry, above all the United States. That is unjust but better than if there were no restrictions. It would be even better if more restriction came about, but that is unlikely. Ironically, some restriction has already come about because of the end of the cold war and the widespread economic crisis in recipient countries. But that restraint is uncodified. It is subservient to the interests in exporting and importing states to continue the institution of the arms trade.

Note

1. Data on the arms trade are notoriously unreliable. There are two sources publishing estimates of worldwide arms trade: the Stockholm International Peace Research Institute (in their annual yearbooks), and the US government (for instance, via the US Arms Control and Disarmament Agency's publication *World Military Expenditures and Arms Transfers*). The data mentioned in the text are based on the latter publication, which purports to cover all arms traded.

Epilogue

A common theme through these chapters is that the occurrence of wars and the killing that they involve are not primarily a consequence of individual aggression. Individual aggression occurs, but it is more a consequence than a cause of war. War motivates aggression and legitimizes killing. The actions of soldiers, sailors and airmen are a consequence of the social situation they are in, a social situation which imposes certain culturally elaborated duties on them. Although combatants usually bear the brunt of war in its most direct sense, their experience in this situation is similar to that of all the ancillary services that make war possible: politicians, munitions workers, doctors and so on have duties specified by the institution of war.

The important message from these chapters therefore concerns the multiplicity of factors that support the institution of war and make war possible. These act in the long term, supporting the idea of war; in the medium term, making weapons available and therefore war possible; and in the immediate context of war, determining the actions of those involved.

In the past, the most effective way to control the incidence of war has sometimes been by the use of superior force (for example, M. Lewis in Chapter 1). The *Pax Britannica* was effective in limited areas in the British colonial era, but the world lacks a central force strong enough and sufficiently motivated to suppress outbreaks of violence worldwide. Somehow, therefore — and in the long run it is a more desirable route — we must undermine the institution of war by countering the factors discussed in this volume. That will be no easy task. People adapt to violence, insecurity and danger, and many of the authors have stressed how militarism or fear of violence may persist in times of peace, and how military images persist in everyday speech, in civilian rituals — and indeed even in churches.

What is certain is that action must be taken on many fronts. It is no use to search for *the* cause of wars. Not only action at the political and economic levels, but changes in cultural attitudes are required. This means not just botching up the present or trying to restore the past, but looking to the future. Kaldor (1993) has argued that the nature of nationalism is changing. The new nationalism, in her view, is out of tune with the times: reacting against the cultural hegemony of earlier nationalisms, it seeks to create ever smaller political units. It also depends heavily on modern media and modern weapons. Kaldor suggests that the new nationalism is non-viable because extensive administrative control over a clearly defined territory is no longer an effective way to manage societies. Instead she advocates greater autonomy at local and regional levels to enhance democracy and foster the diversity of local cultures, to overcome the anonymity induced by modernity and to make locally relevant decisions about local problems. The recognition of ethnic or national difference as a thing to be celebrated, and as a source of pride and identity, is but one side of a familiar coin which tends to have conflict and dominance on account of that same difference as its other face. Perhaps in the late twentieth century, multicultural hybridity may become the end result of a new global world system, but the richness of diversity will be lost. In any case, we do need powerful superordinated organizations able to intervene to protect human rights, uphold environmental and social standards and to prevent war. If we can work towards that goal, more people worldwide may come to share the view that war is not a cruel necessity.

References

Ahern, A. H. (1981) *Chinese Ritual and Politics*, Cambridge: Cambridge University Press

Ahmad, M. (1992) 'Islamic Fundamentalism in South Asia. The Jamat-Islami and the Tablighi-Jamat'. In Marty, M. E. and Appleby, R. S. (eds), q.v.

Aktypis, D. et al. (1990) *Sta Neotera Hronia*, Athens: Orghanismos Ekdhoseon Dhidhaktikon Vivlion

Amherst, H. W. (1947) *H. W. Amherst Papers*, Rhodes House, Oxford

Anderson, B. (1983/1991) *Imagined Communities: Reflections on the Origins and Spread of Nationalism*, London and New York: Verso

Anyon, J. (1979) 'Ideology and US History Textbooks', *Harvard Educational Review,* 49 (3) 361–86

Aran, G. (1992) 'Jewish Zionist Fundamentalism: The Bloc of the Faithful in Israel'. In Marty, M. E. and Appleby, R. S. (eds), q.v.

Arendt, H. (1963) *Eichmann in Jerusalem: A Report on the Banality of Evil*, New York: Viking Press

Armstrong, K. (1988) *The Holy War: The Crusades and their Impact on Today's World*, Basingstoke: Macmillan

Attalides, M. (1979) *Cyprus: Nationalism and International Politics*, New York: St Martin's Press

Attili, G. and Hinde, R. A. (1986) 'Categories of Aggression and their Motivational Heterogeneity', *Ethology and Sociobiology,* (7) 17–27

Ball, N. (1992) *Pressing for Peace: Can Aid Induce Reform?* Washington DC: Overseas Development Council

Barth, F. (1990) *Cosmologies in the Making*, Cambridge: Cambridge University Press

Bateson, G. (1958) *Naven*, Stanford: Stanford University Press

Bauer, W. (1976) Shaw, M. (tr.) *China and the Search for Happiness*, New York: Seabury

Beasley, W. G. (1987) *Japanese Imperialism, 1894–1945*, Oxford: Clarendon Press

—— (1989). In Duus, P., Myers, R. H. and Peattie, M. R. (eds), q.v.

Berghahn, V. R. (1984) *Militarism, The History of an International Debate, 1861–1979*, Cambridge: Cambridge University Press

Best, G. (1979) 'Restraints on War by Land before 1945'. In Howard, M. (ed.) *Restraints on War*, Oxford: Oxford University Press

Bigelow, B. (1991) 'Fighting for the Truth: The Gulf War and our Students'. In Goldberg, S. (ed.), q.v., 50–66

Blacker, C. (1964) *The Japanese Enlightenment*, Cambridge: Cambridge University Press

Blick, J. (1988) 'Genocidal Warfare in Tribal Societies', *Man*, (23) 654–70

Bloch, J. von (1899) *Der Krieg*, Berlin

Boiteau, D., Quinlan, D., Quinlan, P., Stansfield, D. and Stuart-Lyon, G. (1989) *Origins: A History of Canada*, Markham, Ont.: Fitzhenry and Whiteside

Boulding, K. (1979) *Stable Peace*, Austin: University of Texas Press

Bowers. V. and Swanson, D. (1985) *Exploring Canada: Learning from the Past, Looking to the Future*, Vancouver: Douglas and McIntyre (Educational)

Boyd, A. (1987) *Holy War in Belfast*, Belfast: Pretani Press

Brock-Utne, B. (1985) *Educating for Peace: A Feminist Perspective*, Willowdale, Ont.: Pergamon Press

Brodie, M. (1990) *A World Worth Fighting For*, East Wittering, West Sussex: Gooday

Brown, D. (1979) 'The Structure of Poplopa Feasting and Warfare', *Man*, (14) 712–33

Bruce, S. (1987) *God Save Ulster! The Religion and Politics of Paisleyism*, Oxford: Clarendon Press

Brzoska, M. (1994) 'The Financing Factor in the Arms Trade', *Defence and Peace Economics* 7 (1)

Brzoska, M. and Ohlson, T. (eds) (1986) *Arms Production in the Third World*, London: Taylor and Francis

Brzoska, M. and Pearson, F. S. (1994) *Escalation, De-escalation, Negotiation: The Effects of Arms Transfers during Warfare*, Columbus: University of South Carolina Press

Burton, F. (1978) *The Politics of Legitimacy: Struggles in a Belfast Community*, London: Routledge and Kegan Paul

Byas, H. (1944) *Government by Assassination*, London: George Allen and Unwin

Cairns, E. (1987) *Caught in the Crossfire: Children and the Northern Ireland Conflict*, Belfast: Appletree Press

Cameron Commission (1969) *Disturbances in Northern Ireland*, Belfast: HMSO

Canetti, E. (1962) Carol Stewart (tr.) *Crowds and Power*, New York: Continuum

Chagnon, N. (1990) 'Reproduction and Somatic Conflicts of Interest in the Genesis of Violence and Warfare amongst Tribesmen'. In Haas, J. (ed.), q.v., 77–104

Chang, S. H. L. (1990) *History and Legend: Ideas and Images in the Ming Historical Novels*, Ann Arbor: University of Michigan Press

Clanfield, D. (1991) 'Pacijou and the War Toys Project in Quebec'. In Goldberg, S. (ed.), q.v.

Cohn, N. (1976) *Europe's Inner Demons*, St Albans: Paladin

Comaroff, J. L. and Stern, P. C. (1993) *Perspectives on Nationalism and War*, Working Paper, 163, Centre for Studies of Social Change, New School for Social Research, New York

Comstock, G. (1991) *Television and the American Child*, New York: Academic Press

Connerton, P. (1989) *How Societies Remember*, Cambridge: Cambridge University Press

Connor, W. (1994) *Ethnonationalism: the Quest for Understanding*, Princeton: Princeton University Press

Contamine, P. (1984) *War in the Middle Ages*, Oxford and Cambridge, Mass: Basil Blackwell

Cormack, R., Osborne, R. and Gallagher, A. M. (1989) *Religion and Employment in Northern Ireland*, Belfast: Standing Commission on Northern Ireland

Costabile, A., Genta, M. L., Zucchini, E., Smith, P. K. and Harker, R. (1992) 'Attitudes of Parents towards War Play in Young Children', *Early Education and Development*, 3 (4) 356–69

Coufoudakis, V. (1983) 'Cyprus'. In McHale, V. (ed.) *Political Parties in Europe*, Westport: Greenwood Press, 104–33

Dahrendorf, R. (1959) *Homo Sociologicus*, Cologne/Opladen: Westdeutscher Verlag

Daniels, C. (1991) 'Seisan gijitsu iden — seito gijitsu o rei to shite' ('Transfer of Technology for Production — The Case of Sugar'). In Hamashitu Takeshi and Kawakatsu Heitai (eds) *Ajia Koekiken to Nihon Kogyoka, 1500–1800 (The Asian Trade Sphere and Japan's Industrialization)*, Tokyo: Libro

Darby, J. (ed.) (1983) *Northern Ireland: The Background to the Conflict*, Belfast: Appletree Press

Darby, J., Murray, D., Batts, D., Dunn, S., Farren, S. and Harris, J. (1989) *Education and Community in Northern Ireland: Schools Apart?* Coleraine: University of Ulster

Davis, N. Z. (1965) 'The Rites of Violence'. In Davis, N. Z., *Society and Culture in Early Modern France*, Stanford: Stanford University Press

Defence Committee (1990) *The Work of the Defence Committee*. Memorandum submitted by the Chairman of the Defence Committee to the Procedure Committee: Inquiry into the Work of Select Committees, February

Dinges, P. A. and Hitchcock, J. (1992) 'Roman Catholic Traditionalism and Activist Conservatism in the United States'. In Marty, M. E. and Appleby, R. S. (eds), q.v.

Dower, J. (1990) 'The Useful War', *Daedalus*, 119 (3) 49–70

Drucker-Brown, S. (1975) *Ritual Aspects of Mamprusi Kingship*, Cambridge: African Studies Centre, University of Cambridge

Dupuy, T. N. (1980) *The Evolution of Weapons and Warfare*, Fairfax, Va: Hero Books

Duus, P., Myers, R. H. and Peattie, M. R. (eds) (1989) *The Japanese Informal Empire in China*, Princeton: Princeton University Press

Duyvendak, J. J. L. (tr.) (1928) *The Book of Lord Shang*, Chicago: University of Chicago Press

Eisenhower, D. (1961) *Public Papers of the Presidents of the United States: Dwight D. Eisenhower, 1960–91*, Washington DC

Eliasberg, D. (1976) *Le Roman du pourfendeur de démons: traduction annotée et commentaires*, Paris: L'Institut des Hautes Etudes Chinoises

Elisonas, J. (1991) 'The Inseparable Trinity: Japan's Relations with China and Korea'. In Hall, J. W. (ed.) *The Cambridge History of Japan*, vol. 4, Cambridge: Cambridge University Press, 235–300

Ellis, J. (1975) *The Social History of the Machinegun*, New York: Pantheon Books

—— (1980) *The Sharp End of War*, London and Newton Abbot: David and Charles

Elvin, M. (1973) *The Pattern of the Chinese Past*, Stanford: Stanford University Press

Elworthy, S. and Ingram, P. (eds) (1992) 'International Control of the Arms Trade', *Current Decisions Report*, 8, Oxford: Oxford Research Group

Evans, A. S. and Martinello, I. L. (1988) *Canada's Century*, Toronto: McGraw Hill–Ryerson

Eysenck, H. J. and Nias, D. K. B. (1978) *Sex, Violence and the Media*, London: Maurice Temple Smith

Fairbank, J. (ed.) (1968) *The Chinese World Order*, Cambridge, Mass: Harvard University Press

Ferguson, R. B. (1990) 'Explaining War'. In Haas, J. (ed.) q.v. (1990) 26–55

Feshbach, S. (1987) 'Individual Aggression, National Attachment and the Search for Peace: Psychological Perspectives', *Aggressive Behaviour*, (13) 315–25

—— (1990) 'Psychology, Human Violence and the Search for Peace: Issues in Science and Social Values', *Journal of Social Issues*, (76) 183–98

—— (1991) 'Attachment Processes in Adult Political Ideology: Patriotism and Nationalism'. In Gewirtz, J. L. and Kurtines, M. (eds) *Intersections with Attachment*, Hillsdale, NJ: Lawrence Erlbaum, 207–26

—— (1992) 'Parental Attachments, National Attachments, and Attitudes towards War: Some Further Explorations and Thoughts'. In Fraczek, A. and Zumkley, H. (eds) *Socialization and Aggression*, New York: Springer-Verlag

—— (forthcoming) 'Effects of Patriotic and Martial Music on Patriotism and Nationalism'. In Huesmann, R. (ed.) *Edited Volume in Honor of L. Eron*

Feshbach, S., Kandel, E. and Haist, F. (1985) 'Factors Influencing Attitudes towards Nuclear Armament–Disarmament Policies: The Role of Information and the Value Placed on Children'. In Oskamp (ed.) *Applied Social Psychology Annual*, Beverly Hills: Sage Publications, 107–25

Finlayson, J. A. and Haglund, D. G. (1987) 'Whatever Happened to the Resource War?' *Survival*, London: International Institute for Strategic Studies, September/October

Frangoudaki, A. (1979) *Ta Anaghnostika tou Dhimotikou (The Primary School Textbooks)*, Athens: Themelio

Frank, J. D. (1994) 'Some Psychopathological and Sociopsychological Determinants of Bloodthirstiness', *Medicine and War*, (10) 36–49

Fussell, P. (1975) *The Great War and Modern Memory*, London: Oxford University Press

Gallagher, A. M. (1991) *Employment, Unemployment and Religion in Northern Ireland: The Majority Minority Report*, Coleraine: Centre for the Study of Conflict, University of Ulster

Gallagher, J. (1982). In Seal, A. (ed.) *The Decline, Revival and Fall of the British Empire*, Cambridge: Cambridge University Press

Gardner, R. and Heider, K. (1969) *Gardens of War*, New York: Random House

Gellner, E. (1983) *Nations and Nationalism*, Oxford: Basil Blackwell
Gluck, C. (1987) *Japan's Modern Myth*, Princeton: Princeton University Press
Gold, G. (1992) 'Organized Hinduisms: from Vedic Truth to Hindu Nation'. In Marty, M. E. and Appleby, R. S. (eds), q.v.
Goldberg, S. (ed.) (1991) *Times of War and Peace: Dealing with Kids' Concerns*, Willowdale, Ont.: Annick Press
Goody, E. (1991) 'The Learning of Prosocial Behaviour in Small-scale Egalitarian Societies: An Anthropological View'. In Hinde, R. A. and Groebel, J. (eds), q.v., 106–28
Goody, J. R. and Drucker-Brown, S. (1993) 'Babatu', *Anthropology,* 16
Gowing, M. (1988) *Independence and Deterrence*, London: Macmillan
Groebel, J. (1983) 'Aggression and Aggression Research in Germany'. In Goldstein, A. P. and Segall, M. H. (eds) *Aggression in Global Perspective*, New York: Pergamon Press
—— (1992) 'Psychology in Germany'. In Sexton, V. S. and Hogan, J. D. (eds) *Psychology around the World*, Lincoln, Neb.: Nebraska University Press
Groebel, J. and Gleich, U. (1993) *Gewaltprofil des deutschen Fernsehens*, Leverkusen: Leske Verlag & Budrich
Groebel, J. and Hinde, R. A. (1989a) 'The Problem of Aggression'. In Groebel, J. and Hinde, R. A. (eds), *Aggression and War*, Cambridge: Cambridge University Press
—— (eds) (1989b) *Aggression and War*, Cambridge: Cambridge University Press
Haas, J. (ed.) (1990) *The Anthropology of War*, Cambridge: Cambridge University Press
Hardin, R. (1993) 'Self-interest, Group Identity'. In Comaroff, J. L. and Stern, P. C. (eds), q.v.
Harper, D. J. (1982) 'The *Wu-shi-erh ping fang*. Translation and Prolegomena', unpublished Ph.D. thesis, University of California at Berkeley
Harriman, E. (1993) *A Town Called Kozarac*, London: Channel Four Television series: *Dispatches*
Harrison, S. (1993) *The Mask of War*, Manchester: University of Manchester Press
Hastings, M. and Jenkins, J. (1983) *The Battle for the Falklands*, London: Pan Books
Hawkes, D. (tr.) (1985) *The Songs of the South*, Harmondsworth, Middlesex: Penguin
Hegel, R. E. (1985) 'Distinguishing Levels of Audiences for Ming-Ch'ing Vernacular Literature'. In Johnson, D., Nathan, A. J. and Rawski, E. (eds) *Popular Culture in Late Imperial China*, Berkeley: University of California Press, 112–42
Heider, K. (1970) *The Dugum Dani*, Chicago: Aldine Publishing Company
Heilman, S. and Friedman, M. (1992) 'Religious Fundamentalism and Religious Jews: The Case of the Haredim'. In Marty, M. E. and Appleby, R. S. (eds), q.v.
Helsinki Watch (1992) 'War Crimes in Bosnia-Hercegovina', August
Herdt, G. (ed.) (1982) *Rituals of Manhood*, Berkeley: University of California Press

Hinde, R. A. (1989) 'Patriotism: Is Kin Selection both Necessary and Sufficient?', *Politics and the Life Sciences*, (8) 58–61
—— (ed.) (1991) *The Institution of War*, Basingstoke: Macmillan
—— (1993) 'Aggression and War: Individuals, Groups and States'. In Tetlock, P. E., Husbands, J. L., Jervis, R., Stern, P. C. and Tilly, C. (eds) *Behavior, Society and International Conflict*, New York: Oxford University Press, 8–70
Hinde, R. A. and Bateson, P. (1989) 'Some Goals in Education for Peace'. In Hinde, R. A. and Parry, D. A. (eds) *Education for Peace*, Nottingham: Spokesman, 10–17
Hinde, R. A. and Groebel, J. (eds) (1991) *Co-operation and Prosocial Behaviour*, Cambridge: Cambridge University Press
Hobsbawm, E. (1992) 'Ethnicity and Nationalism in Europe Today', *Anthropology Today*, 8 (1)
Hobsbawm, E. and Ranger, T. (eds) (1992) *The Invention of Tradition*, Cambridge: Cambridge University Press
Holmes, R. (1985) *Firing Line*, London: Jonathan Cape
Hourani, A. (1983) *Arabic Thought in the Liberal Ages, 1798–1939*, Cambridge: Cambridge University Press
Howard, M. (1991) *The Lessons of History*, Oxford: Clarendon Press
Howard, Professor Sir Michael (1991) 'Parliamentary Powers, Foreign Policy and National Sovereignty — the Anglo-Saxon Experience', Paper presented to the Netherlands Atlantic Commission Hague Conference on Parliamentary Democracy and International Security Policy, May
Howell, S. and Willis, R. (1989) *Societies at Peace*, London: Routledge
Huesmann, L. R. and Eron, L. D. (1986) *Television and the Aggressive Child. A Cross-national Comparison*, Hillsdale, NJ: Lawrence Erlbaum
Hukanovic, R. (1993) *Deseta Vrata Pakla: Pola Godine Zatocenistva u Logorima Smrti Ornarska i Manjaca* (The Tenth Gate of Hell: Half a Year of Captivity in the Death Camps of Omarska and Manjaca) Oslo: Sypress Forlag
Hulsewe, A. F. P. (1985) *Remnants of Ch'in Law*, Leiden: E. J. Brill
Institute for European Defence and Strategic Studies (1988) *MPs and Defence — A Survey of Parliamentary Knowledge and Opinion,* Occasional Paper No. 36
International Security Information Services (ISIS) (1991) Briefing No. 27, London, May
Iriye, A. (1992) *Japan and China in the Global Setting*, Cambridge, Mass: Harvard University Press
Jackson-Beeck, M and Meadow, R. G. (1979) 'The Triple Agenda of Presidential Debates', *Public Opinion Quarterly,* (43) 173–80
Jansen, M. (1992) *China in the Tokugawa World*, Cambridge, Mass: Harvard University Press
Johnson, G. R. (1986) 'Kin Selection, Socialization and Patriotism', *Politics and the Life Sciences*, (4) 127–54
—— (1989) 'The Role of Kin Recognition Mechanics in Patriotic Socialization: Further Reflections', *Politics and the Life Sciences,* (8) 62–9
Kaldor, M. (1982) *Baroque Arsenal*, London: André Deutsch

—— (1993) 'The New Nationalism in Europe', *Peace Review*, (5) 247–57
Kapferer, B. (1988) *Legends of People, Myths of States*. Washington and London: Smithsonian Institution Press
Kawazoe, S. (1990) 'Japan and East Asia'. In Yamamura, K. (ed.) *The Cambridge History of Japan*, vol. 3, Cambridge: Cambridge University Press, 396–446
Kelly, R. (1968) 'Demographic Pressure and Descent Group Structure in the New Guinea Highlands', *Oceania*, (39) 36–63
—— (1974) 'Political Consciousness and Party Politics in an African Periphery: The Northern Ghanaian Example', unpublished Ph.D. thesis, University of Cambridge
Kelman, H. C. and Hamilton, V. L. (1986) *Crimes of Obedience: Towards a Social Psychology of Authority and Responsibility*, London: Yale University Press
Kim, K. H. (1980) *The Last Phase of the East Asian World Order: Korea, Japan, and the Chinese Empire, 1860–1882*, Berkeley: University of California Press
Kitromilides, P. (1988) 'Koinonikes Shesis ke Nootropies stin Kipro tou Dhekatou Ennatou Aiona', *Ellinika*, (39) 347–80
Knauft, B. (1990) 'Melanesian Warfare: A Theoretical History', *Oceania*, (60) 250–311
Knoblock, J. (tr.) (1990) *Xunzi: A Translation and Study of the Complete Works*, Stanford: Stanford University Press
Koch, K. F. (1974) *War and Peace in Jalemo*, Cambridge, Mass: Harvard University Press
Koliopoulos, J. (1987) *Brigands with a Cause*, Oxford: Clarendon Press
Kosterman, R. and Feshbach, S. (1989) 'Towards a Measure of Patriotic and Nationalistic Attitudes', *Political Psychology*, (10) 257–74
Ladd, G. W. (1992) 'Commentary: Play, Parenting and Peer Partners: Keys to Understanding Children's Social Development?', *Early Education and Development*, 3 (4) 401–6
Ladouceur, P. (1979) *Chiefs and Politicians: the Politics of Regionalism in Northern Ghana*, London: Longman
Lamb, M. (1977) 'The Development of Parental Preferences in the First Two Years of Life', *Sex Roles*, (3) 495–7
Lane, C. (1981) *The Rites of Rulers: Rituals in Industrial Society — The Soviet Case*, Cambridge: Cambridge University Press
Lasswell, H. D. (1927) Propoganda Techniques in the World War, London: Knopf
Lewis, B. (1988) *The Political Language of Islam*, Chicago: Chicago University Press
Lewis, G. (1980) *Day of Shining Red*, Cambridge: Cambridge University Press
Lewis, M. E. (1979) 'The Transformation of the Chinese Military System under the Eastern Han Dynasty', unpublished MA thesis, University of Chicago
—— (1990) *Sanctioned Violence in Early China*. Albany: State University of New York
—— (forthcoming) 'The Feng and Shan Sacrifices of Emperor Wu of the Han'.

In McDermott, J. (ed.) *Court Ritual in China*, Cambridge: Cambridge University Press
Lieven, E. (1989) 'The Psychological Basis of Cooperation and Caring'. In Hinde, R. A. and Parry, D. A. (eds) *Education for Peace*, Nottingham: Spokesman
List, F. (1909) Lloyd, S. (tr.) *The National System of Political Economy*, London: Longmans Green
Loizos, P. (1974) 'The Progress of Greek Nationalism in Cyprus: 1968–1970'. In Davis, J. (ed.) *Choice and Change: Essays in Honour of Lucy Mair*, London: Athlone
—— (1981) *The Heart Grown Bitter: A Chronicle of Cypriot War Refugees*, Cambridge: Cambridge University Press
Loraux, N. (1986) Sheridan, A. (tr.) *The Invention of Athens: The Funeral Oration in the Classical City*, Cambridge, Mass: Harvard University Press
Loza, T. (1993) 'A Civilisation Destroyed', *Balkan War Report*, (20)
Lukes, S. (1974) *Power: A Radical View*, London: Macmillan
Lukovic, P. (1993) 'Media and War: Yugoslavia, the Mirror of Hatred'. In Biserko, S. (ed.) *Yugoslavia: Collapse, War, Crimes*, Belgrade: Centre for Anti-war Action
MacArthur, M. (1977) 'Nutritional Research in Melanesia: a Second Look at the Tsembaga'. In Bayliss Smith, T. and Feachem, R. (eds), *Subsistence and Survival*, London: Academic Press
McCauley, C. (1990) 'Conference Overview'. In Haas, J. (ed.) q.v., 1–25
McKittrick, D. (1993) *The Independent*, London, 21 March
McLean, S. (1986) *How Nuclear Weapons Decisions are Made*, Basingstoke: Macmillan in association with Oxford Research Group
Macmillan, H. (1972) *Pointing the Way*, London: William Collins
McNeill, W. H. (1982) *The Pursuit of Power*, Chicago: University of Chicago Press
Madan, T. N. (1992) 'The Double-Edged Sword: Fundamentalism and the Sikh Religious Tradition'. In Marty, M. E. and Appleby, R. S. (eds), q.v.
Malcolm, N. (1994) *Bosnia: A Short History*, London: Papermac
Mardin, S. (1978) 'Youth and Violence in Turkey', *Archives of European Sociology*, (19) 229–54
Markides, K. (1974) 'Social Change and the Rise and Decline of Social Movements', *American Ethnologist*, 1 (2) 309–30
Marsh, K. and Fraser, J. (1989) *Public Opinion and Nuclear Weapons*, London: Macmillan
Marshall, S. L. A. (1967) *Men Against Fire*, New York: William Morrow and Company
Marty, M. E. and Appleby, R. S. (eds) (1992) *Fundamentalism Observed: The Fundamentalism Project*, vol. 1, Chicago: University of Chicago Press
Meggitt, M. (1977) *Blood is their Argument*, Palo Alto: Mayfield Press
Meznaric, S. and Zlatkovic, J. (1991) 'Gender and Ethnic Violence: The Case of Kosovo', *International Review of Sociology*, (2)
Miall, H. (1987) *Nuclear Weapons: Who's in Charge?* London: Macmillan
Middlebrook (1971) *The First Day at the Somme*, London: Allen Lane and Penguin Books

Millas, H. (1991) 'History Textbooks in Greece and Turkey', *History Workshop Journal*, 22 (2) 21–32

Miyazaki, I. (1959) *Ajiashi Kenkyu (Asiatica: Studies in Oriental History)*, vol. 2, Kyoto: Dohosha

Mizoguchi, T. (1989) 'The Changing Pattern of Sino-Japanese Trade, 1884–1937'. In Duus, P., Myers, R. H. and Peattie, M. R. (eds), q.v.

Moore, P. (1994a) 'Croatia and Bosnia: A Tale of Two Bridges', *Radio Free Europe/Radio Liberty Research Report*, (3) 1, January

—— (1994b) 'A New Stage in the Bosnian Conflict', *Radio Free Europe/ Radio Liberty Research Report*, (3) 9, March

Morton, D. (1988) *Towards Tomorrow: Canada in a Changing World History*, Toronto: Harcourt Brace Jovanovich

Moskos, C. C. (1970) *The American Enlisted Man*, New York: Russell Sage

Mosse, G. (1990) *Fallen Soldiers, Reshaping the Memory of the World Wars*, New York: Oxford University Press

Mottram, R. (1991) 'Options for Change: Process and Prospects', *Royal United Services Institute Journal*, Spring

Murray, D. (1985) *Worlds Apart: Segregated Schools in Northern Ireland*, Belfast: The Appletree Press

Nash, M. (1992) 'Islamic Resurgence in Malaysia and Indonesia'. In Marty, M. E. and Appleby, R. S. (eds), q.v.

Newsam, P. (1989) 'Prejudice and Stereotyping in Relation to Race and Nationality'. In Hinde, R. A. and Parry, D. (eds) *Education for Peace*, Nottingham: Spokesman, 32–41

Nish, I. (1966) *The Anglo-Japanese Alliance*, Oxford: Oxford University Press

—— (1993) *Japan's Struggle with Internationalism*, London: Routledge

Nobel, J. (1991) 'Children as Peacemakers: Child Power'. In Goldberg, S. (ed.), q.v.

Novogrodsky, M. (1991) 'A Toronto Teacher Reflects on Eight Years with FHAO'. In Goldberg, S. (ed.), q.v.

O'Hanlon, M. (1989) *Reading the Skin*, London: British Museum Publications

O'Malley, P. (1993) 'Reinterpreting Northern Ireland', *Register of Research on Northern Ireland*, Coleraine: University of Ulster

Okoye, S. E. (1994) 'Military Expenditures, Equitable Development and Sustainability'. In Smith, P. B. et al. (eds), q.v.

Olweus, D. (1978) *Aggression in the Schools*, New York: Wiley

Osterhammel, J. (1986) 'Semi-Colonialism and Informal Empire in Twentieth Century China: Towards a Framework of Analysis'. In Mommsen, W. J. and Osterhammel, J. (eds) *Imperialism and After: Continuities and Discontinuities*, London: George Allen and Unwin

Oxford Research Group. (June 1991) *Institutional Consequences of a European Security Policy with Particular Reference to Parliamentary Accountability*. Report commissioned by the European Parliament Directorate-General for Research

Papadakis, Y. (1993) 'The Politics of Memory and Forgetting in Cyprus', *Journal of Mediterranean Studies*, 3 (1) 139–54

—— (1994) 'The National Struggle Museums of a Divided City', *Ethnic and Racial Studies*, 400–19

Parsons, J. (1988) *Japan: Its People and Culture*, Edmonton: Reidmore Books
Pick, D. (1993) *War Machines: The Rationalization of Slaughter in the Modern Age*, New Haven and London: Yale University Press
Pierre, A. (1972) *Nuclear Politics*, Oxford: Oxford University Press
Plaks, A. H. (1987) *The Four Masterworks of the Ming Novel*, Princeton: Princeton University Press
Pollis, A. (1979) 'Colonialism and Neo-colonialism: Determinants of Ethnic Conflict in Cyprus'. In Worsley, P. and Kitromilides, P. (eds) *Small States in the Modern World*, Nicosia: New Cyprus Association, 45–80
Prins, G. (ed.) (1983) *Defended to Death*, Harmondsworth, Middlesex: Penguin
Pullinger, S. B. and Higgins, R. (1989) *Parliament and Defence Accountability*. Report based on the work of the Defence Information Group, 1983–87, London
Purcell, H. (1969) *Cyprus*, Great Britain: Ernest Benn
Rabbie, J. M. (1989) 'Group Pressures as Stimulants of Aggression'. In Groebel, J. and Hinde, R. A. (eds) q.v., 141–55
—— (1991) 'Determinants of Instrumental Intra-group Cooperation'. In Hinde, R. A. and Groebel, J. (eds), q.v., 238–62
Ramet, S. P. (1992) *Nationalism and Federalism in Yugoslavia 1962–1991*, Bloomington: Indiana University Press
Ramsbotham, O. (1991) 'Election 1991/1992: The Missing Defence Debate', *Current Decisions Report No 6*, Oxford: Oxford Research Group
Rappaport, R. (1968) *Pigs for the Ancestors*, New Haven and London: Yale University Press
Rheingold, H. L. and Cook, K. V. (1975) 'The Content of Boys' and Girls' Rooms as an Index of Parents' Behaviour', *Child Development*, (46) 459–63
Rokeach, M. (1973) *The Nature of Human Values*, New York: Free Press
Rossabi, M. (ed.) (1983) *China among Equals: The Middle Kingdom and its Neighbours*, Berkeley: University of California Press
Rothschild, K. (1973) 'Military Expenditure, Exports and Growth', *Kyklos*, (26) 4
Sachedina, A. A. (1992) 'Activist Shi'ism in Iran, Iraq and Lebanon'. In Marty, M. E. and Appleby, R. S. (eds), q.v.
Said, E. (1973) *Orientalism: Western Conceptions of the Orient*, London: Pantheon Books
—— (1992) *Culture and Imperialism*, London: Chatto
Sant Cassia, P. (1986) 'Religion, Politics and Ethnicity in Cyprus during the Turkocratia (1571–1878)'. *Archives of European Sociology*, (27) 3–28
Santa-Barbara, J. (1988) 'Learning to Think in New Ways: Rearing and Educating Children to be Citizens in a Peaceful World'. In Ray, D. (ed.) *Peace Education: Canadian and International Perpectives*, London, Ont: Third Eye
Santoni, R. E. (1991) 'Nurturing the Institution of War: Just War Theory's Justifications and Accommodations'. In Hinde, R. A. (ed.), q.v.
Sawyer, R. D. (tr.) (1993) *The Seven Military Classics of Military China*, Boulder: Westview
Scribner, R. W. (1981) *For the Sake of Simple Folk*, Cambridge: Cambridge University Press

Seidel, A. (1987) 'Traces of Han Religion in Funerary Texts Found in Tombs'. In Akizuki Kan'ei (ed.) *Dokyo to Shukyo Bunka*, Tokyo: Hirakawa, 618–714

Serter, V. (1990) *Kibris Tarihi*, Nicosia: Kema Ofset

Shaw, P. R. and Wong, Y. (1989) 'Genetic Seeds of Warfare', *Evolution, Nationalism, and Patriotism*, Boston: Unwin Hyman

Sillitoe, P. (1978) 'Big Men and War in New Guinea', *Man*, (13) 252–71

Singer, J. D. (1989) 'The Political Origins of International War'. In Groebel, J. and Hinde, R. A. (eds), q.v. (1989b), 202–22

Smith, P. B., Okoye, S. E., de Wilde, J., and Deshinkar, P. (eds) (1994) *The World at the Crossroads*, London: Earthscan

Sorabji, C. (1993) 'Ethnic War in Bosnia?', *Radical Philosophy*, (63)

—— (1994) 'Mixed Motives: Islam, Nationalism and Mevluds in an Unstable Yugoslavia'. In Fawzi El-Solh, C. and Mabro, J. (eds) *Muslim Women's Choices: Religious Belief and Social Reality*, Oxford: Berg Press, 108–27

Soysal, M. (1992) 'Political Parties in the Turkish Republic of Northern Cyprus and their Vision of "the Solution"'. In Salem, N. (ed.) *Cyprus: A Regional Problem and its Resolution*, New York: St Martin's Press, 39–43

Spencer, J. (1992) 'Problems in the Analysis of Communal Violence', *Contributions to Indian Sociology*, (26) 2

Srinivasan, A. (1990) 'The Survivor in the Study of Violence'. In Des, V. (ed.) *Mirrors of Violence: Communities, Riots and Survivors in South Asia*, Delhi and Oxford: Oxford University Press

Staniland, M. (1975) *The Lions of Dagbon: Political Change in Northern Ghana*, London: International African Institute

Stern, P. C. (forthcoming) 'Why do People Sacrifice for their Nations?', *Political Psychology*

Stoessinger, J. (1985) *Why Nations Go to War*, New York: St Martin's Press

Strathern, A. (1971) *The Rope of Moka*, Cambridge: Cambridge University Press

—— (1984) *A Line of Power*, London: Tavistock Publications

Strathern, M. (1972) *Women in Between*, London: Seminar Press

Swearer, D. K. (1991) 'Fundamentalistic Movements in Theravada Buddhism'. In Marty, M. E. and Appleby, R. S. (eds), q.v.

Sykes, G. (1989) 'Are We Learning from History?'. In Hinde, R. A. and Parry, D. A. (eds) *Education for Peace*, Nottingham: Spokesman, 63–78

Sykes, S. (1991) 'Sacrifice and the Ideology of War'. In Hinde, R. A. (ed.) q.v., 87–98

Tajfel, H. and Turner, J. (1979) 'An Integrative Theory of Intergroup Conflict'. In Austin, W. G. and Worchel. S. (eds) *The Social Psychology of Intergroup Relations*, Monterey: Brooks/Cole

Tambiah, S. J. (1986) *Sri Lanka: Ethnic Fratricide and the Dismantling of Democracy*, Chicago: University of Chicago Press

Tillyard, E. M. W. (1962) 'A Note on Hate', *Cambridge Review*, 6 June, 358–9

Toby, R. (1984) *State and Diplomacy in Early Modern Japan: Asia in the Development of the Tokugawa Bakufu*, Princeton: Princeton University Press

Van Creveld, M. (1989) *Technology and War*, New York: The Free Press and London: Collier McMillan

Vasic, M. (1993) 'A Bridge too Close, a Peace too Far', *Balkan War Report*, (21)

Wakabasyashi, B. T. (1986) *Anti-Foreignism and Western Learning in Early-Modern Japan: the 'New Theses' of 1825*, Cambridge, Mass: Harvard University Press

Watson, B. (tr.) (1963) *Basic Writings of Mo Tzu, Hsun Tzu, and Han Fei Tzu*, New York: Columbia University Press

Watson, M. W. and Peng, Y. (1992) 'The Relation between Toy Gun Play and Children's Aggressive Behaviour', *Early Education and Development,* (3) 370–89

Webster, R. (1990) *A Brief History of Blasphemy: Liberalism, Censorship and the Satanic Verses*, Ipswich: The Orwell Press

West, R. (1992) 'Determinants of Military Expenditure in Developing Countries: Review of Academic Research'. In Lamb, G. (ed.) *Military Expenditures and Economic Development*, Washington: World Bank

White, H. (1990) 'The Value of Narrativity in the Representation of Reality'. In *The Content of the Form*, Baltimore and London: Johns Hopkins University Press, 1–25

Whitehead, H. (1986) 'Varieties of Fertility Cultism in New Guinea'. *American Ethnologist*, 13 (80–91) 271–89

Williamson, R. (1992) *Profits without Honour? Ethics and the Arms Trade*, London: Council on Christian Approaches to Defence and Disarmament

Wills, J. E. Jr (1984) *Embassies and Illusions, Dutch and Portuguese Envoys to K'ang-hsi, 1666–1687*, Cambridge, Mass: Harvard University Press

Winter, J. M. (1991) 'Causes of War'. In Groebel, J. and Hinde, R. A. (eds) q.v., 194–201

Wurzel, J. (1991) 'The Challenges of Teaching in a Multicultural Environment'. In Goldberg, S. (ed.), q.v.

Yates, R. D. S. (1980) 'The City under Siege: Technology and Organization as seen in the Reconstructed Text of the Military Chapters of Mo-tzu', unpublished Ph.D. thesis, Harvard University

Zeigler, S. and Rosenstein-Manner, M. (1991) 'Bullying at School'. In Goldberg, S. (ed.), q.v.

Index